Praise for *Parallax*

"An accessible and charming history of how we know what we know about our place among the stars." —*Providence Journal*

"Alan Hirshfeld's engaging account moves to a satisfying climax as the exciting race to find stellar parallax heats up. It's a thrilling detective story!"

—Owen Gingerich, Research Professor of Astronomy and the History of Science, Harvard-Smithsonian Center for Astrophysics

"An enthralling, behind-the-scenes glimpse into the lives of . . . heroic stargazers." —*Science News*

"Writing in the genre of Dava Sobel's *Longitude,* Hirshfeld sheds light on the important problem of finding our cosmic place."

—David Levy, science editor, *Parade*

"[*Parallax* shows] historic figures as people, full of false hopes, fear of failure, jealousy, victory, and joy. This comprehensive work . . . will appeal to astronomy buffs and curious readers alike." —*Astronomy*

"Alan Hirshfeld's authoritative and gripping tale of the search for stellar parallax makes me proud to be a part of such a relentlessly curious and persistent species."

—Chet Raymo, columnist, *The Boston Globe,* author of *365 Starry Nights,* and professor of physics and astronomy at Stonehill College in North Easton, Massachusetts

"This book is not just a keeper; it will also become a gift."

—*Sky & Telescope*

"In this thrilling history of the search for parallax, Hirshfeld urges us to 'fly together.' The human failures and frustrations and the crucial discoveries in the quest for parallax are as thrilling as the story of the determination of longitude, and just as important."

—Jane Langton, author of two astronomical mystery novels, *Dark Nantucket Moon* and *The Shortest Day*

"I thoroughly enjoyed reading this excellent book. It is an admirable account, and I'm sure it will be of great value to many readers."

—Sir Patrick Moore, host of the BBC's *The Sky at Night*

"Hirshfeld breathlessly annexes familiar astronomical legends ... and his social history ... engages." —*Publishers Weekly*

"*Parallax* is a fascinating celestial detective story, written in a beautifully lucid, engaging style."

—Dorrit Hoffleit, senior research astronomer, Yale University

"*Parallax*, like Dava Sobel's *Longitude*, is a wonderfully told story of the challenge of measuring our place in the universe. It reaches the stars and the ride is exciting and irresistible."

—Margaret J. Geller, Smithsonian Astrophysical Observatory

"Fun, readable, humorous, gripping, fascinating, sophisticated, informative, and suspenseful all at the same time. And it's about astronomy. Wow." —*Metrowest Daily News*

"With this highly readable and cosmically accessible book, Alan Hirshfeld has done for the measurement of the cosmos what Dava Sobel did for the measurement of longitude. Readers will never again look into the night sky the same way."

—Michael Shermer, publisher, *Skeptic* magazine, author of *Why People Believe Weird Things* and *How We Believe: The Search for God in an Age of Science*

"A delightful history of a crucial advance in knowledge." —*Booklist*

"Alan Hirshfeld has done a magnificent job of telling the story of the race to measure the cosmos. His perspective, in keeping with the best tradition of astronomical history, provides rich insight into the progress and personalities of those who push technology along its cutting edge."

—Leif J. Robinson, editor emeritus, *Sky & Telescope*

Parallax

Parallax

The Race to
Measure the Cosmos

Alan W. Hirshfeld

A W. H. FREEMAN / OWL BOOK
Henry Holt and Company   New York

Henry Holt and Company, LLC
Publishers since 1866
115 West 18th Street
New York, New York 10011

Henry Holt® is a registered trademark of
Henry Holt and Company, LLC.

Library of Congress Cataloging-in-Publication Data

Hirshfeld, Alan.
    Parallax : the race to measure the cosmos / Alan W. Hirshfeld.
        p.   cm.
    Includes bibliographical references and index.
    ISBN 0-8050-7133-4
    1. Parallax—Stars. I. Title.
QB813 .H57   2001
523.8'1—dc21   00-068147

Henry Holt books are available for special
promotions and premiums.
For details contact:
Director, Special Markets.

First published in hardcover in 2001 by W. H. Freeman and Company

First Owl Books Edition 2002

A W. H. Freeman / Owl Book

Designed by Cambraia Fonseca Fernandes

Printed in the United States of America

10 9 8 7 6 5 4 3 2 1

★★★ *To Sasha, Josh, and Gabe,*
the stars of my universe

Contents ★★★

Introduction   xi

Part 1 ★

1   Reinventing the Cosmos   1
2   The Circle Game   19
3   What If the Sun Be Center to the World?   33
4   Crossed Eyes and Wobbling Stars   49

Part 2 ★

5   The heavens Erupt   75
6   The Turbulent Lens   95
7   The Wrangler of Pisa   113
8   The Archimedean Engine   135
9   A Coal Cellar with a View   151
10   Double Vision   171

Part 3 ★

11   Dismal Swamp   193
12   The Twice-Built Telescope   207
13   Quest for Precision   223
14   So Many Grasshoppers   235
15   The Star in the Lyre   245
16   The Subtle Weave   257

Epilogue   A Drink from the Well   269
Acknowledgments   282
Notes and Further Reading   284
Index   305

Introduction ★★★

Meet the photon. Not just any photon. *Your* photon.

What is a photon? The fundamental unit of light: a submicroscopic jot of pure energy, an astronomical Hermes delivering the message of the stars. Nothing appears to distinguish your photon from the trillions of others that breach the surface of the distant star. All are descended from photons born in the star's fiery core. All have survived the turbulent, outbound journey through the star's gaseous envelope. In fact, only one thing sets your photon apart from the others streaking into outer space. This photon is destined to enter your eye. This photon belongs to you.

Your photon never slows along its journey through interstellar space, never strays from its straight-line course. After centuries in the void, it plunges into the Earth's atmosphere, fortuitously avoiding annihilation by air molecules and dusty pollutants. Meanwhile, darkness has fallen on your side of the Earth. Strolling underneath the night sky, you look up. The photon enters your eye, strikes your retina, and turns over its luminous energy to the biochemical process we call "sight." Your photon, plus countless others that precede and trail it, paint in your consciousness the impression of a luminous speck in the heavens. On this clear, dark night, thousands of such specks are visible, altogether a twinkling stellar tapestry arching high above your head. This is the visual siren song that has beckoned observers like yourself since the dawn of humanity. And it is the sight that started me, now some forty years ago, on the road to becoming an astronomer.

From the astronomer's perspective, there is much more to starlight than the visual wallop of a star-studded sky. Bound up in these outer-space photons are clues to the nature of the stars themselves. In the pages that follow, you will read of the astronomers who struggled for centuries to wrest from starlight one fundamental stellar parameter: the distance to a star.

There is no way to determine the distance to a star by a casual inspection of the night sky. With the notable exception of our Sun, stars appear as but luminous pinpoints. Any observable difference in their apparent size stems from the combined distortions of the Earth's atmosphere and the optical instruments—including the eye—through which we observe the heavens. Thus, one might use a star's apparent size to measure the unsteadiness of the air or the optical properties of a telescope, but not to gauge stellar distance. Likewise, a star's brilliance reveals nothing about its remoteness. A visually bright star might be a moderate light-emitter sitting on our solar system's doorstep—or it might be a luminous "supergiant" star parked halfway across the Galaxy. To measure a star's distance requires a pair of human attributes that astronomers have long nurtured: patience and cleverness.

The pathway to the stars is rooted in an everyday phenomenon called *parallax*. Parallax is the apparent shift in an object's position when viewed alternately from different vantage points. Parallax is a primary basis upon which our eyes gauge distance within our surroundings. Distance and parallax go hand in hand: The farther away an object, the smaller the perceived parallax shift. Ancient astronomers had hoped to apply this parallax principle to render distances of celestial objects, such as the Moon, Sun, planets, and stars. Except for the Moon, they were utterly defeated. The cosmos appeared to be far larger than had been supposed.

This book is divided into three parts. The first part chronicles efforts to prove that stars might display a measurable annual "wobble" because of parallax. The linchpin of this assertion is a moving Earth; setting our planet in motion provides it with different vantage points upon the heavens. Thus, stellar-parallax proponents sought to overthrow the traditional model of the cosmos, in which the Earth is central and immobile, and replace it with the revolutionary model of Copernicus (and his predecessor Aristarchus), in which the Earth orbits the Sun. Once that was accomplished, the hunt for stellar parallax—and the laying out of the cosmic third dimension—could begin in earnest.

In the book's second part, we join the initial cadre of astronomers who accepted the most daunting challenge in all of observational astronomy: the detection of a star's parallax. The parallax story is a narrative of failure heaped upon failure, leavened with ample doses of human resilience and unbridled optimism. Generation after generation, astronomers attacked the stellar-parallax problem precisely because it verged on the impossible. These early attempts to measure a star's distance are reminiscent of the assaults on Mount Everest in the 1920s. In both instances, the goal stood plainly in view and the means of conquest were theoretically understood. Both efforts involved crude equipment, naïve assumptions, and rashly optimistic expectations of success. And both foundered because of the sheer magnitude of the ordeal.

The book's third part reveals what finally propelled these Olympian endeavors to their respective goals: planning, persistence, a thorough appreciation of the immensity of the task—and technology. The high-precision telescopes of the early 1800s became to the astronomer what the oxygen canister later became to the mountaineer: the essential ingredient for success.

The quest for stellar parallax is a tale that I—a professional astronomer—thought I knew. But during the eighteen months that I researched and wrote this book, I learned that when one has been schooled by scientists, the explication of scientific theories, techniques, and results eclipses all but the most cursory biographical details. So I write this introduction having completed a journey of discovery. I have come to know a set of extraordinary astronomers in a way I never had. As these astronomers revealed to the world the cosmic third dimension, so too has my research transformed them from flat, iconic caricatures into real people with all the aspirations, emotions, and imperfections that infuse human nature. It was never my intention to knock revered scientists off their pedestals, but rather to invite them to step down and meet me—and you, the reader—on an equal footing. In this book, you will peer over the shoulders of these astronomers as they investigate the heavens. You will have the opportunity to commiserate with them in their crushing failures and revel in their rare successes. You will read of kidnappings, dramatic rescues, swordplay, madness, professional jealousy, hypochondria, and enough angst to fill a universe. Here is a narrative thread that winds it way across the centuries and spins off into the future, a thread that extends from the Earth all the way out to the stars.

[To measure the distance to a star] has been the object of every
astronomer's highest aspirations ever since sidereal astronomy
acquired any degree of precision. But hitherto it has been an
object which, like the fleeting fires that dazzle and mislead the
benighted wanderer, has seemed to suffer the semblance of an
approach only to elude his seizure when apparently just within his
grasp, continually hovering just beyond the limits of his distinct
apprehension, and so leading him on in hopeless, endless, and
exhausting pursuit.
— *John Herschel, to the Royal Astronomical Society, February 12, 1841*

And many strokes, though with little axe, hew down and fell
the hardest-timbered oak.
— *William Shakespeare*, King Henry VI, *Part 3, Act II, Scene 1*

Part 1

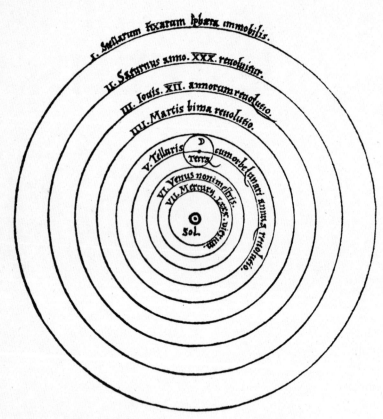

The Sun-centered model of the cosmos. From Copernicus,
De Revolutionibus, 1543.
Source: *Wolbach Library, Harvard University.*

1

Reinventing the Cosmos

The sight of stars always sets me dreaming just as naively as
those black dots on a map set me dreaming of towns and villages.
Why should these points of light in the firmament, I wonder, be
less accessible than the dark ones on the map of France?
—*Vincent van Gogh*, The Letters of Vincent van Gogh

We had the sky, up there, all speckled with stars, and we used
to lay on our backs and look up at them, and discuss about
whether they was made, or only just happened.
—*Mark Twain*, Huckleberry Finn

Free at last! The Earth, wrenched from its central, immobile station in the universe, sailed majestically around the Sun, joining its
sister planets and companion Moon in perpetual motion. This
according to the Greek philosopher-mathematician Aristarchus of
Samos, who proposed the bold rearrangement of the heavens—
replacing central Earth with central Sun—more than two thousand
years ago. After all, he claimed, the Sun—the "lantern" that illu-
minates the heavens—more properly belongs on the throne of the
universe, radiating its light symmetrically over the family of planets. And the Earth, only a fraction of the Sun's size according to
Aristarchus's own calculations, must logically circle the larger
body. Once a year, Aristarchus maintained, our planet completes its
solar circuit, then retraces the identical course again and again, *ad
infinitum.* The other planets—Mercury, Venus, Mars, Jupiter, and
Saturn—likewise move around the Sun; like heavenly fireflies, they
reveal their individual paths against the backdrop of the night sky.

Imagine yourself in Aristarchus's sandals. The year is 270 B.C.
The place is the city of Alexandria at the mouth of the Nile in

northern Egypt. Founded by Alexander the Great in the wake of his campaign of conquest some six decades earlier, Alexandria had grown to become the intellectual and commercial center of the Hellenistic world: a grand galaxy of buildings, monuments, wide ways, and human strivings. Along the boulevardlike Canopic Way, stretching between the Gate of the Sun and the Gate of the Moon, Alexandria's civic vigor manifested itself in spectacular Dionysian processions, one of which "included a hundred-and-eighty-foot golden phallus, two thousand golden-horned bulls, a gold statue of Alexander carried aloft by elephants, and an eighteen-foot statue of Dionysus, wearing a purple cloak and a golden crown of ivy and grapevines."

It was here, after the young Macedonian king's death, that his general, Ptolemy I Soter, established the Temple to the Muses—the "Museum"—and its extraordinary Library with as many as 500,000 documents and scrolls. (By comparison, the largest medieval European library, the Sorbonne, had less than two thousand volumes by the fourteenth century A.D.) A later regent, Ptolemy III, was an even more ardent "book" collector. He decreed that all travelers arriving in Alexandria were to relinquish any documents of literary or scientific value; these were then added to the Library's collection. In return, the travelers got cheap papyrus copies of their "donated" works. Ptolemy once paid a hefty deposit to borrow the "state copies" of Aeschylus, Sophocles, and Euripides from the Athenian library, on the premise of transcribing them; the originals never made it back to Athens.

Alexandria. Centuries of wisdom gathered in one place, a magnet to the world's most able intellects. In addition to its rich Library, the Alexandrian Museum had research rooms, an observatory, a zoo displaying exotic species, living quarters, and a dining hall where scholars gathered to dine and debate. Here was an ancient think tank devoted to the arts and sciences, a precursor Institute for Advanced Study, whose collective scholarship became its legacy to future generations—and whose eventual decline under Christian authority in the fourth century A.D. and destruction in 642 A.D. at the hands of Islamic invaders marked one of civilization's greatest losses. Into this percolating cauldron of ideas, Aristarchus presented his radical theory of the cosmos.

Devising a complete, logical system that explained the movements of the heavenly bodies, the alternation of day and night, the occurrence of eclipses, the uneven lengths of the seasons, the phases of the Moon, and a host of other celestial phenomena was a challenge to the ancients. Their powers of observation were severely limited; they had their eyes and they had their minds, the latter clouded by preconceptions about how the universe *should* be. Physicists Albert Einstein and Leopold Infeld described what these early scientists were up against:

> In our endeavor to understand reality we are somewhat like a man trying to understand the mechanism of a closed watch. He sees the face and the moving hands, even hears its ticking, but he has no way of opening the case. If he is ingenious he may form some picture of a mechanism which could be responsible for all things he observes, but he may never be quite sure his picture is the only one which could explain his observations.

For the three centuries preceding Aristarchus, virtually every Greek philosopher from Pythagoras to Aristotle had hewn to the belief that the *Earth* occupied the hub of the universe and that the Sun, Moon, planets, and even the star-studded celestial sphere that was thought to enclose the universe all circled around it. This "geocentric" mindset had gained inexorable momentum, having barged its way into the bedrock beliefs of generations of deep thinkers by the sheer force of its unchallenged longevity. To these proto-scientists of old, the reality of an Earth-centered cosmos was as self-evident as the contrary "heliocentric" scheme apparently was to Aristarchus.

The geocentric seed germinated around 600 B.C. in the speculations of Thales of Miletus in Asia Minor. Thales, a merchant who made his fortune in olive oil, traveled widely, investigating natural phenomena and conducting experiments. As Plato tells it, Thales once became so entranced by the sight of the stars that he fell into a well while strolling at night. One of his more impressive feats was a measurement of the height of Egypt's Great Pyramid from the length of its shadow. Herodotus also credits Thales with predicting the solar eclipse of 585 B.C., which occurred during a pitched battle between the Medes and the Lydians near the River Halys. The combatants were so terror-stricken to see day turn into night that they called an immediate truce. The eclipse did take place, but the story of its prediction is questionable; probably no one in 600 B.C. knew how to foretell a solar eclipse.

In the arena of cosmology, Thales seems to have taken a cue from his inadvertent dip in the well. He proposed that the universe was constructed around a disk-shaped Earth, which floated serenely in a cosmic ocean. To modern sensibilities, the idea of a water-borne Earth might seem quaint. Yet it marked a turning point in cosmological thought. In previous cosmic models, nothing happened without the intervention of a divine hand. Thales held that his floating Earth formed by *natural* means, akin to the aggregation of silt that he had observed on the Nile delta. The gods might have initiated the Earth's formation, but once the process was up and running, they benignly looked on.

Anaximander, also from Miletus, modified the model proposed by Thales. Anaximander dispensed with the cosmic ocean entirely and suggested that the Earth-disk was free-floating in space, a revolutionary idea at the time. Around 530 B.C., Pythagoras of Samos, whose name adorns the famous theorem relating the sides of a triangle, began to flesh out the crude geocentric model. He proposed that the Earth was not a disk, but rather a stationary *globe,* surrounded by a series of eight concentric, transparent spheres, on which were affixed the Sun, Moon, planets, and stars. The steady rotations of these spheres led to the observed motions of the heavenly bodies, including their daily rising and setting. (Aristotle, that ancient voice of authority who helped enshrine the geocentric model, lent his support to the spherical-Earth hypothesis in his treatise *On the Heavens,* written around 350 B.C. His evidence for a round Earth: the curvature of the Earth's shadow during a lunar eclipse, the changing array of constellations as one journeys north or south, and the disappearance of ships sailing over the horizon. Nonetheless, the Flat-Earth Society persists to this day.)

Planetary motions proved to be more complex than the constant, predictable movement of the stars. Planets sped up and slowed down as they coursed through the sky, and on occasion even looped backwards. Their brightness varied during the year, implying that their distance from the Earth changed. Neither motion variations nor brightness variations were easy to explain in an Earth-centered universe, where each planet supposedly circled the Earth at a fixed distance. Discrepancies between the observed and predicted positions of planets threatened to undo the relatively simple cosmic plan of Pythagoras and his followers. "Observation . . . is the pitiless critic of theory," science historian Agnes Clerke has noted. "[I]t detects weak points, and provokes reforms which may be the beginnings of discovery. Thus, theory and observation mutually act and react, each alternately taking the lead in the endless race of improvement." To survive in the face of the planetary data, the geocentric system had to become more complex.

Around 370 B.C., Eudoxus of Cnidus in Asia Minor, a former student of Plato's and an acquaintance of Aristotle's, raised the heavenly sphere count from eight to twenty-seven. Now each heavenly body had more than one sphere governing its motion. Eudoxus cleverly tilted the spheres at various angles and interlocked their rotations, in order to bring the predictions of his new geocentric model into agreement with the observations. The universe was being transformed into a kind of cosmic machine, a clocklike device that might have sprung from the mind of some divine mathematician.

The ancient Greek astronomers did not necessarily believe in the reality of the cosmic spheres. They viewed them as helpful geometrical constructs, which together formed a virtual computer that emulated the true heavens. As such, it was easy, if need arose, to add yet more spheres and otherwise tinker with the workings of the model, all in the name of reaching accord with the observed motions of the planets. Aristotle himself escalated the number of heavenly spheres to an ungainly fifty-five.

When Alexander the Great conquered Persia around 330 B.C., the extensive astronomical records of the Babylonians fell into Greek hands. These detailed catalogues of planetary positions and solar and lunar eclipses set more stringent requirements on cosmic tinkerers. As a result, new features were folded into the geocentric model. In one modification, the Earth was shoved slightly off-center while the other planets continued to circle the vacated spot. Another enhancement had each planet move around a small orbit, called the *epicycle,* which itself circled the Earth in a wider orbit, the *deferent.* The relative sizes and speeds of all these moving components were adjusted until the model quite effectively simulated what was seen in the sky. The advent of such features eventually reduced the number of cosmic spheres back within reason.

Besides the successful use of the geocentric arrangement of the universe in modeling planetary motions, there were powerful "scientific," emotional, and religious reasons to support such an arrangement. Its aesthetic qualities were pleasing, it conformed to Aristotle's universally accepted idea that dense matter accumulates toward the cosmic center (that is, toward the Earth) while the more evanescent material of the heavenly bodies remains aloft, and it presumably reflected the Creator's obvious intention to construct a humanity-centered cosmos.

Also, one must not disregard the role human nature played in this tale of two theories. Suppose you had spent years designing and building your house. If it's anything like my house, some of the windows don't quite match, not all of the doors shut tightly, and it has to be patched on occasion. But fundamentally it does what a house is supposed to do; it keeps the rain off your head and the winter chill at bay. Now along comes a fellow who takes one look at your house and tells you it's all wrong, dismantle it, rearrange the pieces, and build it over again! No bigger than before, seemingly no better than before, just—well, *different*. How would you react? The geocentrists were hearing essentially the same criticism from Aristarchus about the Earth-centered edifice they had labored to create. It's no surprise that during his lifetime Aristarchus made not a single documented convert within the geocentric camp. Nobody was about to question the old model's fundamental rightness. Nobody, it seemed, except Aristarchus.

Aristarchus's newfangled heliocentric universe, on the other hand, was easy for opponents to criticize. Or simply to ignore.

To the ancient Greeks, the Earth was an impermanent island of life and death, birth and decay, quite distinct from the unalterable perfection of the heavenly realm above their heads. It was simply unfathomable that a base object like the Earth could circle among the higher cosmic spheres, as Aristarchus would have it. As proof, the geocentrists pointed out that at any given moment precisely half of the celestial sphere is visible to Earthbound observers, no matter where they live. Were our planet offset from the cosmic center, observers living on one side of the Earth would see a smaller volume of the universe than those living on the other side.

There was another facet of Aristarchus's theory that disturbed his contemporaries even more: Aristarchus set the Earth spinning. The geocentrists had been taught that the Earth is at rest and that the observed rising and setting of the stars is caused by the continuous turning of the celestial sphere. In the Aristarchian universe, however, the celestial sphere stands motionless, and the nightly progression of the stars from horizon to horizon is revealed for what it is: an illusion. Rotating in an easterly direction, the Earth successively uncovers stars above its eastern limb while eclipsing others behind its western limb. Like a whirling dancer who circles the firelight, our planet revolves about the Sun, spin-

ning all the while—spinning in a twenty-four-hour cycle that creates the ceaseless succession of night and day and the apparent
rising and setting of the stars. (In this suggestion, Aristarchus was
in fact preceded by Heraclides of Pontus, a contemporary of
Aristotle's, who proposed a rotating, albeit geocentric, Earth in
the fourth century B.C.)

The very concept of a spinning Earth was preposterous to the
geocentrists, who appealed to common experience. Stand outside
at night, they suggested, and watch the stars drift slowly across the
sky; there is absolutely no feeling of motion underfoot. Why deny
the credibility of one's own senses? Obviously, the celestial sphere
was in motion, not the Earth. That the sphere of the stars was
spinning didn't faze the geocentrists in the least; the heavenly
realm wasn't subject to the same kinds of physical restrictions to
which the Earth was. Furthermore, by the third century B.C. the
geocentrists already had a rough idea of the Earth's size. If our
planet's surface truly completed a circuit in a mere day's time,
they argued, continents would hurtle around the Earth's axis at
hundreds of miles an hour. Gale-force winds would perpetually
rake the landscape. Oceans would inundate low-lying cities.
Anything—or anyone—not firmly anchored to the ground would
tumble away into the sky. How could Aristarchus explain the
absence of such cataclysmic effects?

Or take a more prosaic example. Toss a stone straight
upward and watch it inevitably return to the thrower's hand.
Were the Earth spinning, the geocentrists reasoned, the stone
would veer backward from its starting point, left behind while
the Earth sweeps the thrower forward. Only in retrospect is the
counterargument clear: The ancients didn't understand the concept of inertia. It would be many centuries before Galileo
explained why a stone dropped from the mast of a moving ship
strikes the deck directly below its starting point, not several feet
back. The stone maintains the forward inertia it acquired from
the ship's motion through the water. The same is true for a stone
tossed upward from the ground, only this time the role of the
moving ship is played by the Earth itself. The thrower, the stone,
even the air that envelops them all move in unison with the
Earth's rotating surface. The stone maintains this "sideways"
velocity—that is, it keeps up with the moving thrower—even
while rising or falling through the air. So a rotating Earth is not
counter to observation. Then again, neither was the stationary
Earth of the geocentrists.

The stars themselves posed a special challenge to Aristarchus's heliocentric theory, for now the geocentrists put forth a stringent observational test. Were the Earth truly in orbit, at times it would swing closer to stars on a given side of the celestial sphere. As a result, stars would vary in brightness during the course of the year, brighter when the Earth was nearer, fainter when it was farther away. For example, during winter (or, specifically, winter in the northern hemisphere), when the Earth supposedly swings toward the constellation Canis Major, the prominent star Sirius should brighten; months later, when the Earth is moving away from that part of the sky, Sirius should dim. But Sirius displays no such annual variation in its light. Nor does any other star in the heavens.

While the stars were constant, the planets did vary periodically in brightness. Aristarchus could have easily explained why. In a heliocentric cosmos, planet-to-planet separations change all the time as these bodies make their individual orbital rounds; when a planet swings closer to the Earth's current position, it appears more prominent in the sky. At the time, the geocentric model was unable to account for changing planetary brightness because each of its planets maintains a constant distance from the central, fixed Earth.

Were the Earth in orbit, there would be yet another observational consequence: parallax. Parallax (from the Greek *parallassein*, "to change") is the apparent shift in position of an object when viewed alternately from different vantage points. Parallax manifests itself in our everyday lives because each of our eyes has its own distinct perspective on the surrounding world. Our eyes are separated by a couple of inches, and that's enough to make an object's position appear somewhat different to each eye. As you read the words on this page, for example, your eyes inevitably cross a little, a physical consequence of the parallax effect. Except for very nearby objects, such as a finger held right in front of your nose, this convergence of your eyes is imperceptible. Yet our brain has evolved the ability to interpret the degree of eye-crossing as a gauge of an object's parallax, and hence its distance. The farther the object, the smaller its parallax.

The words on this page induce a fair degree of eye-crossing; they are close by and consequently have a relatively large parallax. A plant across the room displays a smaller parallax, and a tree down the street a smaller one yet. What about a star? Does a star far off in space exhibit a parallax? The answer depends on whom you ask. The geocentrists would tell you that stars show no parallax, that with only a few inches separating our eyes—the twin vantage points from which we observe the heavens—both eyes effectively see the star in the identical position.

But what if the vantage points were more widely separated? (Sounds painful, but don't worry, it requires no rearrangement of your face.) Suppose you sighted the star from Greece and had a compatriot do the same from, say, Egypt, many hundreds of miles away. Would the star's position appear different to observers at these two locations? Would the star display a parallax? Yes, the geocentrists would answer, *in principle* the star might show a parallax, given the substantial distance between the vantage points. But the geocentrists would hasten to point out that there is absolutely no *observed* parallax of stars when viewed from these two locations or indeed from locations even farther apart. Evidently, the celestial sphere is so remote that a pair of observers standing at *opposite ends* of the Earth—the widest possible "baseline" in a geocentric universe—would never detect a star's parallax.

Now let's pose the parallax question to Aristarchus, the heliocentrist. In *his* universe, there is a much higher expectation of observing stellar parallax, for the Earth's wide-ranging movement around the Sun provides vantage points whose separation is ever so much larger than the Earth itself. An astronomer viewing stars over such a broad baseline—the width of the Earth's *orbit*—might be able to detect a perceptible change in the position of a star. In fact, during the course of the year, the star should appear to sway to and fro, perhaps even dance a little oval against the firmament, a skewed reflection of our planet's orbital motion around the Sun.

Yet to the pretelescopic vision of the ancients, not a single star danced Aristarchus's parallactic "jig" in the heavens. The stars just marched monotonously up from the eastern horizon and down to the western horizon, with not a twist or a twirl among them. Once again, the evidence weighed against Aristarchus's assertion that the Earth revolved about the Sun. Too bad, for the eventual key to plumbing the stellar depths did lie in just that parallactic wobble that Aristarchus presumably had forecast but would never live to see. Today we know that even the most prominent stellar parallax falls far below the threshold of naked-eye vision. But nobody knew that back then.

The moth-eaten fabric of recorded history contains not a jot about Aristarchus's response to his critics. Only one of his written works survives, a short treatise on the sizes and distances of the

Sun and Moon, and it doesn't even mention the heliocentric universe (although it does have implications for the model). For all we know, Aristarchus might have been dismissed as a crank, a fringe thinker whose ideas were too far removed from the mainstream to be taken seriously. Nevertheless, in the spirit of supporting the underdog, let's take up the debate on Aristarchus's behalf. The primary issue at hand: stellar parallax. We will infer Aristarchus's response to the parallax problem—the inexplicable constancy of the stars in the face of a moving Earth—from a single sentence in an unusual work by one of Aristarchus's contemporaries: Archimedes.

Archimedes was born around 280 B.C. in Syracuse, principal city-state of Sicily. His interests ranged widely across mathematics, physics, astronomy, and engineering. Although Archimedes himself valued his theoretical research most, he managed to develop a host of practical inventions, including an arsenal of devastatingly effective engines of war used to defend Syracuse from Roman invaders. Here is an account of what hostile armies faced when going up against the terrifying weaponry of Archimedes:

> When the Roman legions attacked, they were met with a rain of missiles and immense stones launched from giant catapults. . . . Trying to protect themselves under a cover of shields, the helpless Roman infantry was crushed by boulders and large timbers dropped from cranes that swung out over the battlements. Most horrific of all were the enormous clawlike devices that wrecked the Roman fleet as it tried to enter the harbor, shaking the ships about and even plucking them clean out of the water.

As Plutarch tells it, "such terror had seized upon the Romans, that, if they did but see a little rope or a piece of wood from the wall, instantly crying out, that there it was again, Archimedes was about to let fly some engine at them, they turned their backs and fled." The Roman general, Marcellus, could do nothing but settle in for a long siege.

In the nonmilitary realm, Archimedes developed the water screw that bears his name: an inclined, hand-turned, hollow spiral that is still used in underdeveloped countries to raise water from irrigation ditches. Archimedes also experimented with levers, pulleys, and fulcrums. "Give me where to stand," he was said to have boasted to the Syracusan king, Hieron, "and I will move the Earth." To prove his point, Archimedes rigged up a device with which he single-handedly launched a fully loaded ship from dry dock. Afterward Hieron declared that "Archimedes was to be believed in everything he might say." As for Archimedes, he was content to return to his theoretical pursuits; his machines had been designed "not as matters of any importance, but as mere amusements in geometry."

Archimedes displayed the eccentricities often associated with genius. In his *Lives,* Plutarch describes how Archimedes would "forget his food and neglect his person, to that degree that when he was occasionally carried by absolute violence to bathe or have his body anointed, he used to trace geometrical figures in the ashes of the fire, and diagrams in the oil on his body, being in a state of entire preoccupation, and, in the truest sense, divine possession with his love and delight in science."

The most frequently told tale about Archimedes has to do with a royal crown. King Hieron had contracted with a local goldsmith to fashion for the king a new crown out of solid gold. The crown was duly completed, and it looked magnificent. But the king had heard a rumor that the craftsman had squirreled away some of the gold that had been allocated for the crown, substituting an equal weight of a lesser metal like silver or copper. (Nowadays, jewelers routinely mix gold with other metals for strength; common fourteen-karat gold is only 58 percent pure.) Hieron asked Archimedes to assess the "purity" of the new crown.

Gold is pretty heavy stuff: A three-inch gold cube tips the scales at nearly nineteen pounds. (Talk about the burdens of high office; no wonder George Washington refused to wear royal trappings.) Archimedes conceived a plan. First he would weigh Hieron's crown. Then he would measure the crown's volume. Finally he would check whether the crown's weight was proper for that volume of gold. But like all crowns, this one was irregularly shaped. How was Archimedes to determine its volume?

According to Plutarch, Archimedes was mulling over this royal conundrum one day while at the public baths. As usual, the water rose as Archimedes settled into the tub. But that day this commonplace phenomenon rang with new significance. His own body had pushed aside the volume of water that used to occupy the space where he now lay. The water had risen in proportion to the volume of his body. Archimedes realized how he might measure the volume of Hieron's crown. He'd fill a tub to the brim with water, immerse the crown, and capture the overflow in a measuring cup. The crown's volume would simply equal the volume of water in the cup. Upon receiving this flash of insight, Archimedes is alleged to have leaped from his bath and run home naked through the streets of Syracuse, shouting that now-famous expletive of scientific discovery, *"Eureka!"* ("I have found it!") He determined the crown's volume and found that it was indeed lighter than the equivalent volume of gold. Archimedes reported to King Hieron that the crown had been "cut" with silver. The goldsmith was executed.

This alleged streaker of bygone days, Archimedes, played an indirect but critical role in the story of stellar parallax. Having spent part of his life in Alexandria concurrently with Aristarchus, Archimedes was familiar with the revolutionary heliocentric system. Like everyone else at the time, he didn't believe it. But there was one instance late in his life where Aristarchus's outlandish idea proved useful to Archimedes, a fact that he recorded. Had he not, we might never have known any details of Aristarchus's heliocentric hypothesis.

Around 216 B.C., shortly before his death at the hands of a Roman soldier, the septuagenarian Archimedes completed a mathematical work for King Hieron's successor, Gelon II. This curious tract, written for the layperson and entitled *The Sand-Reckoner,* was an attempt by Archimedes to demonstrate his facility with the mathematics of large numbers. To wit, Archimedes set out to compute the number of sand grains required to fill the universe! His pitch has the definite ring of the carnival barker: "I will try to show . . . that of the numbers named by me . . . some exceed not only the number of the mass of sand equal in magnitude to the Earth filled up in the way described, but also that of a mass equal in magnitude to the Universe." Sounds impressive, even today. But just how big was this universe that Archimedes intended to "fill"?

To make a lasting impression on his royal patron, Archimedes wanted a universe vaster than the commonly accepted geocentric system, with its tightly nested heavenly spheres. So without passing judgment on the relative merits of the two competing world systems, he selected Aristarchus's alternative model, which he knew to be sizable indeed. It is through Archimedes that we gain insight into the *dimensions* of the new heliocentric universe by way of Aristarchus's own words, as set down in *The Sand-Reckoner.*

First let's lay the groundwork for what Aristarchus had to say about the extent of the universe, for he approached the issue in an oblique and somewhat problematic fashion. In general, how can one effectively convey the size of an object? One method is to state its dimensions outright: The house is thirty feet wide. Another method is to compare its size to that of another object or, equivalently, to form the size *ratio* between the objects: The shopping mall is twice the length of a football field, or the ratio between the length of the shopping mall and the length of a football field is 2. Yet a third method is to express the size ratio of two objects in terms of the size ratio of two other objects: The tree's height is to the man's height as the man's height is to the child's height. In this case, if the man is six feet tall and the child three feet tall, the ratio of their heights is 2; because the tree's height bears this same ratio to the man's height, the tree is twelve feet tall. It is the last of these methods that Aristarchus chose to reveal the size of his universe. Here, in modernized form, is what Archimedes claims Aristarchus wrote:

> The distance of the stars bears the same relation to the diameter of the Earth's orbit as the surface of a sphere bears to its center.

To my ear, Aristarchus's statement is eerily reminiscent of one of those thorny problems on the SAT. Parsing it, we see that the statement contains two ratios: The first ratio is a purely physical one, between the *distance of the stars* and the *diameter of the Earth's orbit;* the second ratio is a purely mathematical one, between the *surface of a sphere* and the *center of a sphere.* (Aristarchus might have been referring to the celestial sphere, but any sphere will do.) These two ratios, Aristarchus claimed, are equal, which we can represent as follows:

(distance of the stars) / (diameter of the Earth's orbit) =
(surface of a sphere) / (center of a sphere)

Compute the second ratio, the mathematical one, and you simultaneously have learned the first ratio, the physical one. Here Aristarchus has apparently offered Archimedes a way to compute the extent of the heliocentric universe ("distance of the stars") in terms of the diameter of the Earth's orbit. Specifically, how many times bigger is the universe compared to the Earth's orbit? That we can answer by computing the second ratio. Or can we?

The geometrically savvy reader will already have recognized that the second ratio, between the *surface of a sphere* and the *center of a sphere,* is patently absurd. Yet similar phrasing was often used by Greek astronomers when expressing the enormity of the heavens. (By "surface of a sphere," we'll take Aristarchus's meaning to be the sphere's diameter.) The center of a sphere is a point, which geometrically speaking has no size at all. Thus, a sphere's diameter is *infinitely* larger than its center, because any number divided by zero yields infinity. In other words, the second ratio turns out to be infinite. Taken literally, Aristarchus's statement implies that the stars are infinitely far away. An infinite universe did not suit Archimedes, who at this point might have identified with the hapless protagonist in physicist George Gamow's limerick:

> There was a young fellow from Trinity
> Who took [the square root of infinity]
>   But the number of digits
>   Gave him the fidgets;
> he dropped Math and took up Divinity.

In *The Sand-Reckoner,* Archimedes wished to perform a calculation that involved the purported extent of the universe. He was unable to deal with one that was infinitely large, so he "fudged" his assumptions a bit. He chose a less literal interpretation of Aristarchus's enigmatic phrase comparing the surface to the center of a sphere. According to Archimedes, Aristarchus must have meant something like this:

> *The distance of the stars bears the same relation to the diameter of the Earth's orbit as the diameter of the Earth's orbit bears to the diameter of the Earth.*

Or, in ratio form:

> (distance of the stars) / (diameter of the Earth's orbit) =
> (diameter of the Earth's orbit) / (diameter of the Earth)

Voilá! The second ratio is no longer infinite, but a straightforward comparison between two measurable quantities: the *diameter of the Earth's orbit* and the *diameter of the Earth.* These were numbers for which crude estimates already existed (from Aristotle, among others). Using generous assumptions, Archimedes reasoned that the diameter of the Earth's orbit did not exceed 10,000 Earth-diameters. (In reality, he was still too low by a factor of 2.) The previous relation becomes:

(distance of the stars) / (10,000 Earth-diameters)
= (10,000 Earth-diameters) / (1 Earth-diameter)

Therefore, according to Archimedes, the stars on the celestial sphere are 100 million Earth-diameters away, which dwarfs the estimate of 10,000 Earth-diameters proposed by the geocentrists. (Yet it still falls far short of the true distance to even the *nearest* star: over 3 *billion* Earth-diameters.) Next Archimedes converted his cosmic distance to *stadia,* a terrestrial measure used by the ancients and roughly equivalent to a tenth of a mile. For this calculation, he "fudged" his assumptions yet again. Remember, Archimedes was trying to impress King Gelon by "filling" the *biggest* possible universe, so he unabashedly goosed up the numbers for maximum effect. He took the Earth's diameter to be nearly 1 million stadia, which he freely admitted was tenfold larger than the commonly accepted value at the time. In the end, Archimedes deduced that the Aristarchian universe was about 100 trillion stadia in radius, roughly 10 trillion miles. To fill it required 1,000 trillion trillion trillion trillion trillion sand grains. I imagine that King Gelon was suitably impressed. (In modern "scientific notation," the number of sand grains would be written 10^{63}, that is, 1 followed by sixty-three zeroes.)

Archimedes died shortly after completing *The Sand-Reckoner,* slain by a soldier when the invading army of the Roman general Marcellus surprised drunken sentries and swept into Syracuse. According to Plutarch, the soldier came upon Archimedes scratching geometric shapes in the dirt and became incensed when the old man scolded him for interrupting. More than a century later, in 75 B.C., Cicero restored the neglected tomb of Archimedes, having discovered it by the legendary figure carved on its face: a cylinder enclosing a sphere. The tomb is now lost.

Regardless of how Archimedes chose to interpret Aristarchus's original statement about the size of the universe, Aristarchus's intent is clear. He sought to explode the geocentrists' notion of the compact universe, whose not-so-distant celestial sphere coddled humanity like a cosmic eggshell. The celestial sphere, in his view, was enormous compared to the breadth of the Earth's orbit. The stars are exceedingly far away. And *that,* we may infer, is how Aristarchus would have responded to his critics on the stellar parallax issue: Stars exhibit no discernible change in their light and no measurable parallax because they are so remote. If a star is greater than a million Earth-diameters away, its parallax will be undetectable by the naked eye. The celestial sphere in the Aristarchian universe lay far beyond that threshold, so there was absolutely no chance for the ancients to witness stellar parallax. In the debate between the two world systems, Aristarchus stood on weak ground, for the absence of stellar parallax confirmed nothing one way or the other. Either the Earth was in motion and the stars were immensely distant or the Earth was fixed and central and the stars were relatively nearby. There was simply no way to distinguish an undetectable parallax from a nonexistent parallax.

The geocentrists surely railed against the immense void that Aristarchus had imposed between the celestial sphere and the inner realm of the Sun and planets. From their perspective, there could be no considerable gap among the spheres occupied by the heavenly bodies; each sphere brushed up against its neighbors like the layers of an onion. Why would the Creator have built a universe with so much wasted space? It just didn't make sense.

The Sun-centered cosmos, as conceived by Aristarchus, had no real chance of acceptance within the Hellenistic academy. The geocentric model was firmly rooted by this time and accounted well enough for the observed motions of the Sun, Moon, and planets. In hindsight, we know that the geocentric model was ultimately discredited; yet we shouldn't judge its ancient proponents too harshly. Whatever their religious or philosophical biases, their arguments appealed to a most sensitive human touchstone: common sense. To believe the heliocentric picture of the heavens, even today, means distrusting one's own senses, for in truth the concept of a moving, spinning Earth is counterintuitive. Despite decades as an astronomer, I still find it hard to fathom that the Earth streaks through half a million miles of space while we snooze away the night in our beds. Prior to the 1700s, there simply was no definitive proof of the Earth's orbital motion. The heliocentric system of Aristarchus was also largely qualitative, its mathematical underpinnings apparently not worked out. So while in principle it could account for the observed lunar and planetary movements, no one at the time knew for sure. And most important, the heliocentric system lacked *unambiguous* observational evidence, such as stellar parallax.

In the eyes of his fellow philosophers, Aristarchus was at best a radical thinker, at worst a heretic. At least one tract denouncing him for impiety was published during his lifetime, a bilious indictment by the Stoic philosopher Cleanthes of Assus, suitably titled

Against Aristarchus. The heliocentric model apparently gained only one major adherent, the mathematician Seleucus of Seliucia on the Tigris River, and that was a full century after the new theory had been proposed. No, the time was not yet right for an unbiased hearing of such revolutionary ideas. Aristarchus stood as much chance of reversing the geocentric juggernaut as a mosquito has of turning back a freight train.

Had stellar parallax been observable by the ancient Greeks, the case for the Sun-centered universe might have been won in Aristarchus's time. But as the situation played out, the clockwork universe continued to tick for another 1,800 years before the parallax issue emerged once more in a renewed conflict between the two competing world systems. And yet another 300 years passed until, in a nearly simultaneous burst of competitive energy, three astronomers vied to pin down the long-sought parallax.

The Earth-centered model of the cosmos. From Hevelius, *Selenographia*, 1647.
Source: *Wolbach Library, Harvard University.*

2

The Circle Game

It isn't that they can't see the solution. It is that
they can't see the problem.
—*G. K. Chesterton*

The history of cosmic theories . . . may without exaggeration
be called a history of collective obsessions and
controlled schizophrenias.
—*Arthur Koestler*, The Sleepwalkers

Step outside for a night of sky-gazing and you might at first believe that astronomers would have little trouble coming up with an accurate and coherent model of planetary motion. Planets shine with a brilliant, steady light, and once someone points them out to you in the sky, you'll find it easy to distinguish the planets from their twinkling stellar counterparts. As the hours tick by, the planets seem to match the uncomplicated, regular motion of the stars, which travel steadily upward from the eastern horizon, then downward over the western horizon. A simple, uniform, twenty-four-hour cycle. But as days, weeks, and months pass, you'd come to appreciate the origin of the word *planet:* It's Greek for "wanderer." And wander they do. While patently nonuniform over the long term, the motions of the planets are not so irregular as to be perverse. No, a plan clearly underlies the movements of these heavenly wanderers, but a plan of maddening subtleties. To ancient astronomers, it must have seemed as though the Creator were intent on confounding their attempts to explain planetary motion.

The Sun, too, behaves somewhat like a wandering planet. One month it might appear in the constellation Virgo, yet some weeks later it will have crept into the constellation Libra. (Obviously, the constellations are not visible when the Sun is up; the ancients used the Moon's position among the stars to infer which constellation the Sun occupied.) The Sun's movement against the starry background is gradual—on average about 1 degree a day—but its motion accelerates during northern-hemisphere autumn and winter, then slows down during spring and summer. The *year* is defined as the time it takes the Sun to complete one such circuit around the heavens. The Sun's annual path is not parallel to the diurnal movement of the stars, but inclined by about 23 degrees. Of course, this solar march against the starry backdrop is illusory; the Earth's orbital motion makes the Sun *appear* to creep through the constellations. But the ancients, with the exception of Aristarchus and a follower or two, didn't believe that.

The planets' heavenly tracks are generally aligned with the solar path, but their movements are unlike the Sun's. Mercury and Venus never stray far from the Sun in the sky; the two planets jog back and forth, sometimes leading, sometimes trailing the Sun in its motion. Neither is visible in the glare of day nor in the midst of night, but both show up either at dawn as a morning "star" or at dusk as an evening "star." The remaining planets progress at varying speeds eastward through the constellations. And, at intervals, each also brightens and executes a temporary reversal in direction: a retrograde loop to the west.

The ancients never believed that the planets actually halted in space and traveled backward for a while; they assumed there was a mechanism by which the motion *appeared* retrograde from our vantage point. They also believed in the Aristotelian ideal that planets move with constant speed in circular orbits. Therein lay the seemingly insurmountable challenge to astronomical model-makers: how to account for a planet's observed irregular movements without violating the Aristotelian principle of circular motion at constant speed. That these model-makers nearly succeeded is a testament to their ingenuity.

Aristarchus didn't fare well in the planetary-motion arena. After he passed from the scene around 230 B.C., his heliocentric hypothesis languished. Like the proverbial Detroit concept car, it sported a sleek, tradition-breaking design—but there was nothing "under the hood." Neither Aristarchus nor his proximate successors had fitted the heliocentric model with the engine of mathematical refinement. As such, it could not achieve the gold standard of ancient astronomy: the accurate prediction of solar, lunar, and planetary positions. The heliocentric hypothesis might have been intriguing, but without predictive power, it simply wasn't practical.

The geocentric model's history followed a very different arc. In the second century B.C., the preeminent observer of antiquity, Hipparchus, unqualifiedly cast his lot with the Earth-centered theory of the heavens. Little is known of Hipparchus's life beyond his astronomical observations. At various times, he lived in Nicaea in northern Turkey, in Rhodes at the eastern end of the Aegean Sea, and perhaps in Alexandria. He died after 127 B.C., the date of his last recorded observation. As with Aristarchus, only one of Hipparchus's original treatises survives, and it gives no measure of the breadth or significance of his work.

Hipparchus was the first astronomer to truly appreciate the intimate link between theory and observation, how they cross-check one another to reveal flaws and contradictions in our knowledge. Unlike Aristarchus a century before, Hipparchus was not so much interested in method for method's sake, but primarily as it pertained to—and in the end accounted for—real phenomena in the sky. That's why Hipparchus was such a tireless observer. He knew that without the nourishment of superior data, theories are little more than educated conjectures about nature—compelling perhaps, yet not anchored to reality. A theory without supporting facts is like a half-formed sculpture; it might appear promising, but one must await its completion before judging its worth.

Besides measuring the positions of the planets far more accurately than anyone had before, Hipparchus may have compiled the first extensive star catalogue. The catalogue—whose existence is based solely on reports by later astronomers—would have been considered an utter waste of effort by some of Hipparchus's contemporaries: The stars were fixed relative to one another, and their light never varied, so what's the point of enumerating this information? However, a remarkable celestial occurrence might have stimulated Hipparchus to assemble his catalogue. In 134 B.C., the familiar, glittery patterns of the constellations changed overnight; a "new" star had appeared in the sky. This was an unprecedented event, and it seemed to contradict the prevailing wisdom about the immutability of the heavens. We don't know whether the new star was a nova—a dim, hot star that occasionally erupts—or perhaps a tailless comet. It appeared, shone for a time, then faded away. (If it was, in fact, a star, it already existed when Hipparchus noticed it but was too dim to be seen before-hand; the actual "birth" of a star is not an overnight process but one that lasts many human lifetimes.) The historian Pliny, in his *Natural History*, gives this account of the new star:

> *Hipparchus . . . detected a new star that came into existence in his lifetime; the movement of this star in its line of radiance led him to wonder whether this was a frequent occurrence, whether the stars that we think to be fixed are also in motion; and consequently he did a bold thing, that would be reprehensible even for God—he dared to schedule the stars for posterity, and tick off the heavenly bodies by name in a list, devising machinery by means of which to indicate their several positions and magnitudes [brightnesses], in order that from that time onward it might be possible easily to discern not only whether stars perish and are born, but whether some are in transit and in motion, and also whether they increase or decrease in magnitude—thus bequeathing the heavens as a legacy to all mankind, supposing anybody had been found to claim that inheritance.*

Nineteen centuries later, English astronomer Edmond Halley, for whom the famous comet is named, claimed Hipparchus's "inheritance." Halley compared the current positions of stars against those recorded in an ancient catalogue that may have been based on Hipparchus's and proved that some stars indeed exhibit individual motions through the heavens. In compiling data for his own stellar catalogue, Hipparchus, too, analyzed a sample of stars that had been studied in previous centuries. Strangely, all of the stars in his sample had shifted from their former positions in the sky. The shifts were not random, as were those later uncovered by Halley, but rather were systematic. It was as though the starry celestial sphere had tipped slightly, relative to the Earth, thereby altering the coordinates of every star.

The astronomer's sky-coordinate system is based on the Earth's orientation with respect to the stars; any change in that orientation will cause a change in the coordinates of all heavenly bodies. By Hipparchus's reckoning, the positions of celestial bodies were shifting *en masse* by about 1 degree every century. He realized that any sky coordinates recorded by his predecessors must be corrected—shifted to *his* era—before they could be compared with his own observations. In this way, Hipparchus gained a centuries-long record of solar, lunar, and planetary movements, which were gauged against the positions of stars. It would be yet another eighteen centuries before Isaac Newton explained the mass drifting of celestial coordinates over time. The Earth's axis, he knew, lists into the ocean of space like a windblown sailboat. Under the gravitational tug of the Moon, our planet's axis gyrates, or *precesses,* in the manner of a child's top. The precession of the Earth's axis is gradual, taking all of 26,000 years to complete one circuit. Precession sweeps the Earth-linked sky-coordinate system past the stars, causing star positions to change over time.

Until Hipparchus's time, the geocentric model's predictive capability, although far better than its heliocentric competitor's, was still uncomfortably approximate. Hipparchus's crucial discovery of precession permitted astronomers for the first time to accurately piece together current observations with those of the distant past. Armed with his own and his many predecessors' data, newly unified through precession adjustments, Hipparchus revised the geocentric model's mathematical elements to bring the predictions for the Sun and the Moon into closer accord with reality. Hipparchus didn't live long enough to revamp the entire geocentric system; the motions of the five known planets had not been addressed. Someone else would have to finish the job. The stage was set for one of the most influential figures in the history of science, an individual whose writings dominated astronomical thought for 1,400 years. Enter Claudius Ptolemy.

Ptolemy (not to be confused with Egypt's royal Ptolemys) lived in Alexandria during the second century A.D. Everything we know about him—which is precious little—has been gleaned from later commentaries or from his surviving works. Here is a man who wrote with the supreme confidence of one who had heard the *Word* revealed, the Word in this instance being the entire plan of the physical universe. The biblical allusion is not unwarranted; for fourteen centuries, Ptolemy's writings commanded almost religious allegiance. Ultimately, his model of the universe became part of Christian doctrine. In the eyes of most, the "Ptolemaic" universe *was* the universe.

Ptolemy embodies the culmination of ancient Hellenistic science. He was amazingly efficient at transforming astronomical ideas and research into compelling reading. His master work, the thirteen-volume *Megale Syntaxis,* or "Great Compilation," was completed around 150 A.D. Through the ages, it has become better known by the laudatory name bestowed upon it by a ninth-century

Arabic translator: the *Almagest*—the "Greatest." Here is amassed a mind-numbing array of numerical tables, technical diagrams, mathematical proofs, step-by-step examples, and detailed explanations: altogether a sophisticated primer on how to compute the coordinates of the Sun, Moon, and planets for all time. Not to short-change the stellar universe, Ptolemy tossed in an up-to-date star catalogue with more than a thousand entries. Were the *Almagest* a work of music, it would not be just a Beethoven symphony; it would be *all* of his symphonies, with detailed instructions on how they should be played and on how you, too, could write a symphony.

Of all the surviving mathematical treatises from antiquity, only Euclid's *Elements* rivals the *Almagest* in scope, complexity, and intellectual force. The structural similarity between the two works is not accidental; Ptolemy, like Aristarchus before him, adopted the *Elements* as the model for his volume. The *Elements* and the *Almagest* are equally ambitious: One claims to encompass the whole of geometry; the other, the whole of astronomy. Astronomer Owen Gingerich writes: "Ptolemy's attempt to write the astronomical equivalent of Euclid's geometry was all too successful: nobody bothered to copy any competing treatises that might have been available."

The centerpiece of the *Almagest* is Ptolemy's version of the geocentric model of the heavens. The "Ptolemaic" model, as it became known, is the geocentric model on steroids—bulked up with mathematical sophistication and unsurpassed predictive capability. Ptolemy used past observations to generate custom geometric models for the Sun, the Moon, and each of the planets. From this he assembled numerical tables that his readers could use to compute each object's future path in the sky. His system functioned like a computer—relevant data went in, planetary positions came out—but the user, not the computer, performed all the tedious calculations. Like a master teacher, Ptolemy patiently led his readers through examples of the entire procedure. The *Almagest* was to be *used,* not just studied.

One issue Ptolemy was forced to deal with was the patently nonuniform motions of certain celestial bodies. The Sun, for example, appears to move through the sky with varying speed depending on the time of year. Planets, too, appear to speed up and slow down as the months pass. But according to the writings of Aristotle, whose word was law in those days, heavenly bodies, in fact, move always at constant speed in circular orbits. How then could Ptolemy make the uniformly moving Sun and planets *appear* to move nonuniformly in the sky? He accomplished this feat with a geometric sleight of hand introduced three centuries earlier by Apollonius of Perga and subsequently refined by Hipparchus: Ptolemy shoved the Sun's orbit sideways such that it was no longer centered on the Earth. (The Earth remained the center of the celestial sphere, so the universe was still geocentric.) As a result, the "eccentric" Sun was at times closer to the Earth and at other times farther away. When the Sun was closer to the Earth, it would appear to move faster in the sky, and when it was farther from the Earth, it would appear to move more slowly. (To use a terrestrial analogy, imagine yourself—the "Earth"—standing on the infield of a circular race track, not at the center but close to one side. A race car—the "Sun"—moves around the track at constant speed. Yet from your offset position, the car *appears* to move more quickly when it rounds the close-by curve than when it rounds the curve far across the infield.) Ptolemy offset the Sun's orbit by the proper amount and in the proper direction to reproduce the Sun's varying movement in the sky. At no point did Ptolemy's eccentric model violate Aristotle's tenet of uniform circular motion.

The motions of the other heavenly wanderers, however, were more complex than the Sun's and demanded models with more than just an eccentric adjustment. For instance, how was Ptolemy to generate a retrograde loop within a planet's path when Aristotle did not permit any alteration of the planet's speed, much less a complete reversal in its direction? In effect, Aristotle was Ptolemy's King Eurystheus, demanding that he perform the cogitative version of one of the Herculean labors. Ptolemy found a partial solution in the concept of epicycles and deferents, which, like the eccentric orbit, was the brainchild of Apollonius of Perga.

Ptolemy held that every planet circles around a small orbit, the *epicycle,* whose center simultaneously swings around a larger orbit, the *deferent,* which itself is centered on the fixed Earth. This also happens to be the basis of a dizzying amusement park ride I once encountered called the Twizzler. (In a previous life, it was probably used by NASA to winnow out irresolute astronauts.) The Twizzler consists of four metal arms that extend radially from a motorized central hub. Each arm supports four benches arranged symmetrically in a circle. When the machine is running, the benches whirl in a tight orbit about the end of the arm—the epicycle—while the arm sweeps the benches around in a large orbit—the deferent. Seen from above, the cumulative motion of each bench is a rosettelike pattern. As each bench swings through the innermost part of its epicycle, it appears from above to momentarily stand still or even reverse direction. Or, viewed from the Twizzler's central hub—analogous to the Earth's position in Ptolemy's universe—the bench, or Ptolemy's "planet," appears to generally move forward but occasionally executes a retrograde jog.

Here's how it happens. The accompanying illustration represents a bird's-eye view of the Twizzler. The bench moves through its epicycle at constant speed, indicated by the arrow labeled v; simultaneously the entire epicycle circles the deferent at constant speed V. At any given moment, the bench's overall speed is a combination of its epicycle and deferent speeds. How do we properly combine the epicycle and deferent speeds to arrive at the bench's overall speed? The deferent speed V is always in the "forward" direction (here, counterclockwise). But the bench's epicycle speed

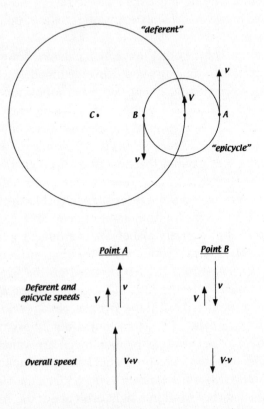

An epicycle-based amusement park ride.

v changes direction all the time: sometimes it's directed forward, as at point **A**; sometimes backward, as at point **B**.

The bench's overall speed at any given moment is found by combining the epicycle and deferent speeds. Thus, the overall speed of the bench at point **A** is **V** + **v**, because at this instant both the epicycle and deferent speeds are directed forward. At point **B**, the bench's overall speed is the difference **V** − **v**, because here the epicycle and deferent speeds are oppositely directed; as the deferent motion tries to sweep the bench forward, the epicycle motion tries to sweep it back.

What would the bench's motion look like from our vantage point at the Twizzler's hub, the Earth's position in the Ptolemaic system? Specifically, how would the bench appear to move when rounding the inner part of its epicycle near point **B**? If the epicycle and deferent speeds are equal, they cancel each other out and the bench will appear to stand motionless for an instant. Or if the epicycle speed is *greater* than the deferent speed, the bench appears to jog backward. Retrograde motion!

Ptolemy's system was a mathematical version of the Twizzler ride that simulated the motions of the planets. By adjusting the input data—relative sizes of epicycle and deferent, relative epicycle and deferent speeds, tilts of epicycle and deferent in space—Ptolemy could generate an overall motion that nearly matched the planet's observed motion in the heavens, retrograde loops and all. The *Almagest* portrayed the process in exquisite detail for the Moon and each of the five known planets. It was a triumph of mathematical astronomy. But for Ptolemy it was not good enough.

Even after extensive adjustment of epicycles and deferents, there remained differences between the predicted and the observed positions of the planets. The differences must have gnawed at Ptolemy's sensibilities, for in the *Almagest* he takes an unprecedented and startling step to erase them. Adapting his eccentric-Sun model, Ptolemy nudges aside each planet's large deferent orbit such that the orbit's center no longer coincides with the Earth, but with a vacant spot away from the Earth, labeled **C** in the accompanying illustration. Now the epicycle's distance from the Earth changes as it moves along the deferent. So far, nothing controversial. Next Ptolemy conjures up another point in space, essentially an "anti-Earth," equally removed from the center **C** as is our planet, but diametrically opposite. This point Ptolemy calls the *equant*. The epicycle's distance from the equant

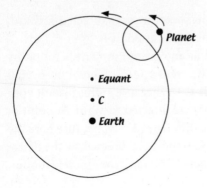

Ptolemy's equant concept.

likewise changes, as it did previously with the Earth, but in the opposite sense: When the epicycle is close to the equant, it is far from the Earth, and vice versa.

Now Ptolemy takes the fateful step that would raise the hackles of astronomers for centuries to come. In Book IX of the *Almagest,* Ptolemy posits that the rate at which each planet's epicycle moves along its deferent is mathematically governed by the placement of the equant. It's as though some invisible arm extends across space from the equant and sweeps the epicycle around, not uniformly, but at a speed that varies: faster when the epicycle is far from the equant, slower when the epicycle is close. The speed variation is such that the epicycle *appears* to move uniformly to a hypothetical observer at the equant. To an observer on the Earth, however, the speed variation *is* evident as the planet speeds and slows along its night-sky path. With his equant model, Ptolemy stood in stark defiance of Aristotle's bedrock dictum of uniform circular motion in the heavens. Precisely why the planetary models were more accurate with the equant refinement wouldn't become clear until the 1600s when Johannes Kepler concluded that planets move in ellipses, not circles. At the time, Ptolemy only knew that his models worked better once Aristotle's restriction of uniform speed had been cast aside. And for Ptolemy, predictive accuracy outweighed all other considerations. Even the wisdom of the great sage himself.

The *Almagest* was a how-to manual of planetary motion. Each planet was dealt with in turn, separately from the others. In a later treatise called *Planetary Hypotheses,* Ptolemy assembled the planets into a coherent whole. No longer was his brainchild just a mathematical construct to predict planetary positions but an entire physical universe with real bodies traversing real orbits: the Ptolemaic system. Ptolemy arrayed the heavenly bodies around the Earth in logical order, depending on how quickly they moved through the sky: Moon, Mercury, Venus, Sun, Mars, Jupiter, and Saturn. He assumed that each planet's epicycle and deferent circled within the hollow of its own thick, transparent crystalline sphere and that these planetary spheres nestled around one another with no intervening gaps. Encasing the whole arrangement was the sphere of fixed stars. Ptolemy even wrote down the dimensions of each planetary sphere and of the universe as a whole: Starting at the center, one could lay almost 10,000 Earths side by side before reaching the celestial sphere.

Despite their misgivings about Ptolemy's insistence on nonuniform motion, astronomers were confident that the story of the universe had now essentially been told. In a sense, Ptolemy was the Bill Gates of his day. His Ptolemaic "operating system," despite its known deficiencies, grew to dominate—in fact, monopolize—the astronomical marketplace. It became the sole paradigm through which the universe was considered. And with the Earth planted motionless at its hub, held fast by Ptolemy's mathematical persuasiveness, stellar parallax would be absent from astronomical discussion for as long as the Ptolemaic system endured.

After Ptolemy came a steady decline in the quality of Western scientific activity. Centuries of domination by Rome followed by Christendom's rejection of "pagan science" reduced the intellectual hub of Alexandria to a shadow of its former self. The great Library went up in smoke during the Islamic invasions of the seventh century, and with it much of the fruit of the Hellenistic philosophers. Europe became an intellectual backwater, having lost virtually all of the significant documents of antiquity. Few European scholars knew how to read Greek. Here and there, fragments of the ancient texts could be found, most translated inaccurately into Latin. Only meager portions of Euclid's and Aristotle's writings were available. Ptolemy and his remarkable system of the heavens were virtually forgotten. In effect, the Hellenistic learning tradition had evaporated from the consciousness of the continent's inhabitants, as they struggled to survive the economic and political hardships of the day. Starting with the disintegration of the Roman empire in the fifth century A.D. and lasting the better part of a millennium, Europe wallowed in the Dark Ages.

At the same time, Islamic science flourished. Scientists in the Islamic world, which by the seventh century stretched throughout the Mediterranean region, had inherited the ancient manuscripts that the Christian West had lost. All of the major Hellenistic treatises were translated into Arabic, to which Islamic scientists added their own observations and reflections. It was here that Ptolemy's work was preserved for future generations.

In the tenth century, Europe began its gradual emergence from the Dark Ages. With growing political stability, improving economic conditions, and the reinvigoration of commerce came a renewed interest in learning. Trade provided more contact with the Islamic world, and the works of antiquity were rediscovered. Arabic translations of the ancient texts were rendered into Latin and disseminated throughout Europe. During the twelfth and thirteenth centuries, universities were established, at first to study the ancient texts, but later to promote original investigation and creative work. Latinized versions of Ptolemy's *Almagest* and Aristotle's treatises became part of the standard university curriculum.

"When a teacher in a Faculty of Arts first laid hands on the *Almagest*," writes historian Michael Hoskin, "it must have represented a level of technical virtuosity exceeding by several orders of magnitude any astronomical writings he had previously known." Imagine that same medieval scholar running a finger down the *Almagest*'s mathematical tables, gazing at Ptolemy's mysterious epicyclic diagrams, pondering their implications. Could it truly be that such an extraordinary work sprang from a long-ago human mind? Was it possible that the mechanics of the universe had been understood a thousand years earlier? Experiencing the *Almagest* would have been as revelatory to the medieval scholar as one of Einstein's books on relativity would have been to Ben Franklin. With the work of Italian Dominican friar Thomas Aquinas in the thirteenth century, the Ptolemaic cosmos began to acquire a sacred dimension and was ultimately assimilated—with "suitable" theologically driven modifications—into Christian doctrine.

In the 1400s, maritime exploration and the attendant engine of commerce fueled a resurgence of interest in astronomy. Portugal's Henry the Navigator erected an observatory in 1420 on the headland of Sagres at the extreme southwest point of Europe. For the next forty years, he ran an institute devoted to astronomy, the navigational arts, and maritime exploration.

Astronomy in the fifteenth century meant Ptolemy's astronomy: the geocentric universe. Its predicted planetary positions could be several degrees away from reality, but with the observational arts in such disarray, even errors of this magnitude were tolerated. But Ptolemy's grip on the universe could not last forever. The Earth would eventually shake free from its conceived central position, for the seed of the Ptolemaic system's demise was planted within its very structure. And in sixteenth-century Poland, under the careful tending of a single astronomer, the seed began to sprout.

The Rittenhouse orrery.
Source: *Denise Applewhite/Princeton University Office of Communications.*

3

What If the Sun Be Center to the World?

> . . .What if the Sun
> Be Center to the World, and other Stars
> By his attractive virtue and their own
> Incited, dance about him various rounds?
> —*John Milton,* Paradise Lost, *Book VIII, 1697*

> A rock pile ceases to be a rock pile the moment a single man
> contemplates it, bearing within him the image of a cathedral.
> —*Antoine de Saint-Exupéry,* Flight to Arras, *1942*

On display in the lobby of Princeton University's astronomy building is an orrery completed in 1770 by Pennsylvania clockmaker and self-taught astronomer David Rittenhouse. An orrery is a dollhouse universe, a mechanized model of the heavens whose clockwork gears drive miniature celestial bodies in synchrony with their real counterparts in the night sky. The orrery derives its name from Britain's Charles Boyle, Earl of Orrery, for whom the first such device was built in the early 1700s.

I was an undergraduate at Princeton from 1969 to 1973, one of only four astronomy majors in a class of almost a thousand. Every day, on my way to or from the basement cubicle in Peyton Hall where I studied, I'd pass the Rittenhouse orrery. Even in today's high-tech, visually assaultive world, the orrery's gilded orbs and star-speckled face grab one's attention. The device is four-foot square and set vertically in a wall off the building's main lobby. Its surface is polished sheet-brass, painted navy blue. An axle projecting from the center of the orrery's face bears a brass ball representing the Sun. Around it are arrayed six smaller balls—the known planets circa 1770, including our Earth—which revolve in scaled-down elliptical orbits, properly shaped and correctly inclined relative to one another, "their velocities so accurately adjusted," according to Rittenhouse, "as not to differ sensibly from the tables of Astronomy in some thousands of years." Each planet is accompanied by its retinue of satellites, also orbiting at the proper pace. The entire mechanism used to run off a wind-up spring but is now driven by an electric motor.

Surrounding the orrery's face is a motorized brass ring engraved with celestial coordinates, such that the progress of each planet through the heavens can be read directly. The ring rotates 1 degree every seventy-two years, to correct the coordinate readings for the slow gyration of the Earth's axis. The Rittenhouse orrery is a mechanical time machine: Turn a winch and you will see how the solar system appeared up to 5,000 years in the past or how it will appear up to 5,000 years in the future. Historical eclipses and eclipses yet to come can be observed and their dates recorded.

The Rittenhouse orrery had been intended for the College of Philadelphia, now the University of Pennsylvania. But on April 23, 1770, shortly before the orrery was completed, President Witherspoon of Princeton University—or the College of New Jersey, as it was then known—handed Rittenhouse £300 and claimed the device. John Adams came to see it for himself in August 1776, and British soldiers played with it during the Revolution. (The soldiers broke the orrery, but it was repaired after the war.) The orrery astonished spectators at the 1893 World's Fair in Chicago, then disappeared until 1948 when it was found, still crated from the fair, in the basement of a Princeton building. The orrery was restored in 1952 and mounted in Princeton's Firestone Library before being moved to its current home in the lobby of Peyton Hall.

Nearly thirty years have passed since I stood in that lobby, bleary-eyed from studying, just to gaze at the wondrous blue dial. It seemed almost too marvelous to believe—and still seems so now—that the heavenly movements could be so accurately reproduced by a 200-year-old assemblage of gears, springs, and cams. At the time David Rittenhouse constructed his diminutive clockwork universe, the long-running argument about the arrangement of the solar system's bodies had ceased. In university lecture halls around the world where astronomers had taught the geocentric theory only two centuries before, they now presented the helio-centric theory instead. Today in the lobby of Peyton Hall one can see in shining splendor the same heliocentric system envisioned thousands of years ago by Aristarchus. Yet even in Aristarchus's day, the heliocentric system had been spurned in favor of its geo-centric competitor. So why is it that the big brass globe of the Sun, and not the tiny orb of the Earth, sits at the orrery's hub? How did astronomers come to be so confident of the heliocentric arrange-ment that David Rittenhouse would devote three years of his life to celebrate it in brass?

On October 12, 1492, after a ten-week voyage across the Atlantic, Christopher Columbus set foot in the Americas. Half a world away in Kraków, a nineteen-year-old college student named Nicolaus Copernicus was preparing a voyage of his own: a voyage that would span the entirety of his adult life, a voyage not of sail but of mind.

The world in which Copernicus grew up was exploding with new ideas. Although it predates our so-called Information Age by half a millennium— and although the only

Nicolaus Copernicus. From an engraving dated 1682.
Source: *Owen Gingerich.*

"computers" at the time were very human ones hunched over columns of figures—Copernicus's era has equal claim to the "Information Age" title. Since Johann Gutenberg's development of the movable-type press in the mid-1400s, approximately 40,000 printed volumes had swept across Europe in a Renaissance version of the Internet. A number of these books fell into Copernicus's capable hands. (During the Thirty-Years War, Copernicus's entire book collection was carted off to Sweden, where it presently resides in the library of Uppsala University. Theft, to be sure, but at least we know what Copernicus was reading.)

Through the printed page, the voices of the ancient philosophers spoke to the many who hungered for enlightenment. Scholars communicated with one another freely, exchanging ideas that earlier might have subjected them to ridicule or worse. Artists like Leonardo, Michelangelo, and Raphael depicted in their works flesh-and-blood human figures—*people,* not emotionally vacant caricatures. Nature once again became a frontier to be explored instead of feared. "The lurking demons that inspired terror in the hearts of medieval men," according to writer Rudolph Thiel, "had vanished like puffs of smoke. The whole of creation appeared fresh as on the seventh day, waiting only to be discovered, enjoyed, depicted, and explained."

Nicolaus Copernicus was born in 1473 in the Polish city of Toruń on the banks of the Vistula River. Orphaned at age ten, Copernicus nonetheless led a privileged life, insulated from the economic deprivation and the day-to-day hardships of his less fortunate countrymen. Upon the death of his father, a prosperous merchant, Copernicus (along with his brother and two sisters) fell under the guardianship of his maternal uncle, Lucas Watzenrode. Described as a "tornado of a man," the ruthless Watzenrode clawed his way up the Church hierarchy until his installation in 1489 as bishop and ruler of the small episcopal duchy of Warmia (known in the German-speaking countries as Ermland). This Baltic-coast enclave served as a periodic battleground in the ongoing enmity between the Polish king and the rapacious Teutonic Knights, a religious military order dating back to the time of the Crusades. Throughout his tumultuous life, Bishop Watzenrode unabashedly used his position to advance his beloved nephew. To Nicolaus Copernicus, Uncle Lucas was a surrogate father.

Kraków's Jagiellonian University, where the bishop enrolled Copernicus, was noted throughout Europe for its academic excellence and open-mindedness. Students were exposed not only to the classic Greek works but also to those of the Islamic world. It was the only university on the European continent with two astronomers on its faculty, and judging from the surviving inventory of books in Copernicus's personal library, the Jagiellonian astronomers must have seen much of the young man.

At the university, Copernicus learned mathematics and Euclidean geometry, the physics of Aristotle, and of course the Ptolemaic system of the universe, although not from the *Almagest* itself but from a fifteenth-century summary in Latin by Viennese astronomer Georg Peurbach. He also would have heard the views of dissenting Greek philosophers who had proposed a spinning Earth or, as ridiculous as it might have sounded at first, an Earth that moves through space. It's likely that the Jagiellonian professors nurtured the seeds of doubt in Copernicus's mind about Ptolemy's model.

In 1496, Bishop Watzenrode arranged for nephew Nicolaus to continue his studies at the bishop's own alma mater, the University of Bologna. After trekking across the Alps to Italy with his brother, Andreas, Copernicus took up the study of canon law to prepare for an ecclesiastical career. But at the same time, he indulged in his foremost passion: astronomy. Copernicus rented quarters from the noted Italian astronomer Domenico Maria da Novara, who instructed him in the art of astronomical measurement. Novara was quite candid with his assistant about his frustrations with the Ptolemaic system. Perhaps someday a young man such as Copernicus would come along and fix it.

I imagine that Copernicus's heart beat faster on the night of March 9, 1497, for Nature had provided him a "blessing in the skies": a direct test of Ptolemy's model. At 11 o'clock that evening, the Moon passed in front of the bright star Aldebaran in the constellation Taurus. Copernicus certainly understood the significance of such an eclipse: By timing the interval between the disappearance and subsequent reappearance of the star, he could accurately compute the Moon's diameter. According to Ptolemy, the Moon swung in a wide epicycle as it circled the Earth; sometimes the Moon was closer to the Earth and sometimes farther away. Therefore, the Moon's diameter should appear to vary as it cycles through its phases. Yet no obvious variation had ever been observed. Could there be perhaps a more subtle variation? In timing the eclipse of Aldebaran, Copernicus concluded that the half-Moon appears precisely the same diameter as the full Moon. Under the skies of Bologna in 1497, Copernicus saw with his own eyes the fallibility of the Ptolemaic system.

Evidently Copernicus gained some measure of recognition as an astronomer while in Italy, for he delivered an astronomical lecture in Rome during the jubilee year 1500. Meanwhile, Bishop Watzenrode had arranged a lifetime appointment for Copernicus back home in Warmia: administrative canon at the cathedral in Frombork (or Frauenburg, in German). The position carried minimal responsibilities and a comfortable stipend generated by revenues from Church properties. Copernicus returned to Poland barely long enough to unpack his bags. In 1501, he headed back to Italy to study medicine at the University of Padua (which, during the 1590s, would number among its faculty a young professor named Galileo). Copernicus remained in Padua for two years, except for a brief period at the University of Ferrara, where he obtained his doctorate in canon law. He refused to be ordained as a priest, even when threatened with suspension of his stipend.

In 1506, the thirty-three-year-old Copernicus returned to Warmia for good. He'd enjoyed fifteen years of the finest university education and was fluent in astronomy, mathematics, economics, medicine, canon law, and several languages, including Greek. He'd even dabbled in art. In short, Copernicus appears to have come back from Italy the archetypal Renaissance man. He moved in with his uncle at the bishop's palace at Lidzbark and became Watzenrode's trusted personal secretary, physician, and legal advisor—all the while receiving the stipend for his no-show job in Frombork. Over the next decade, Copernicus traveled regularly on medical consultations, diplomatic missions, and political errands for his uncle.

In 1512, Bishop Watzenrode died suddenly after attending Polish King Sigismund's wedding feast in Kraków. Rumors abounded that the bishop had been poisoned by agents of his long-time foe, the Teutonic Knights. Now nearly forty years old and robbed of his uncle's beneficence, Copernicus finally took up his post in Frombork, which he once described as the "most remote corner of the Earth." He moved into a secluded three-story tower along the city's northwest wall, where he would live out the rest of his life. In 1513, Copernicus purchased "800 building stones and a barrel of lime" to raise his own *turricula*, or observing platform. From here he had an unobstructed view of the sky, although the rising mists from the Frisches Haff lagoon made astronomical observations difficult at best. Not that it made much difference to Copernicus. Since returning from Italy, his energies had been directed inward to the development of his radical theory of the cosmos.

As potent as the Ptolemaic system was, in Copernicus's opinion it exuded an air of arbitrariness, as though it had been cobbled together from disparate elements:

> *[Past astronomers] have not been able to discover or to infer the chief points at all—the structure of the universe and the true symmetry of its parts. But they are just like someone*

*taking from different places hands, feet, head, and the other
limbs, no doubt depicted very well but not modeled from the
same body and not matching one another—so that such parts
would produce a monster rather than a man.*

Copernicus objected on philosophical grounds to Ptolemy's
equant-point concept and the nonuniform planetary motion that
it implied. When it came to planetary motion, Copernicus hewed
unswervingly to the Aristotelian party line that all heavenly bodies move in circles at constant speed. There was a practical objection, too. If Ptolemy's planetary arrangement were physically real,
as many medieval scholars believed, any mechanical linkage
between the equant point and a planet's epicycle and deferent
would necessarily poke through the intervening crystalline
spheres. How could a moving linkage slice freely through solid
material? (Ptolemy never fully addressed this point.) Lastly,
Copernicus felt that Ptolemy had not satisfactorily explained the
various irregularities in the length of the year, as defined by the
changing position of the Sun in the heavens.

Sometime between 1506 and 1514—a period during which
Michelangelo painted a swirling universe on the Sistine ceiling,
Balboa sighted the Pacific, and Machiavelli wrote *The Prince*—
Copernicus penned a sentence of almost poetic simplicity that
would nonetheless reverberate through history: "What appear to
us as motions of the Sun arise not from its motion but from the
motions of the Earth." The sentence is part of a brief treatise
known today as the *Commentariolus,* or "Little Commentary." The
Commentariolus was Copernicus's opening salvo against the
Ptolemaic system, which he declared "neither sufficiently perfect
nor pleasing to the mind." He outlines his heliocentric model in
seven postulates, then concludes with the promise that a full
mathematical treatment would be forthcoming in a more extensive work.

It was only in 1515, after the release of the *Commentariolus,*
that a complete version of Ptolemy's *Almagest* became available.
The original text had been translated from its original Greek into
Arabic, then from Arabic into Latin, yet it was the best available to
Copernicus. Leafing through its pages, Copernicus must have
realized what an immense task lay ahead of him. To supplant such
an edifice would require a work of comparable magnitude and
mathematical rigor. Copernicus would have to tackle in grueling
detail every one of Ptolemy's claims. In effect, he'd have to create

his own *Almagest* for the heliocentric system. Copernicus was forty-two years old; his work would not end until he was nearly seventy.

The reticent Copernicus distributed handwritten copies of his *Commentariolus* to only a small circle of scholars. As a result, word of its revolutionary content spread snail-like through the wider world. But eventually, like a chain reaction, interest swelled until the learned clamored to know more about the new theory—and about the mysterious canon who had proposed it. Papal secretary John Albert Widmanstadt presented Copernicus's ideas to Pope Clement VII in 1533. At least one high-ranking Church official, with papal approval, urged Copernicus to publish a complete version of his thesis. The Church was unperturbed by Copernicus's radical ideas; in the eyes of the Church hierarchy, the heliocentric model was a mathematical fiction: a tool for predicting celestial phenomena and not a literal rendering of the universe.

Word began to spread of the solitary canon in Frombork who had devised a new theory of the heavens. German astronomer Erasmus Reinhold, a future convert to the heliocentric system, mentioned in a commentary, "I know of a modern scientist who is exceptionally skillful. He has raised a lively expectancy in everybody. One hopes that he will restore astronomy." Among those who heard such reports was Erasmus Reinhold's colleague, Georg Joachim Rheticus, an enthusiastic twenty-five-year-old mathematics professor at the University of Wittenberg. Too eager to wait for Copernicus to publish his book, Rheticus set out in the spring of 1539 for Frombork to read the handwritten manuscript itself. He bore a suitable gift for the now-elderly astronomer: newly published, pigskin-bound editions of Euclid and Ptolemy—in Greek.

If only Rheticus's biography of Copernicus had survived the centuries, we might have thrilled to the narrative of his first encounter with Copernicus: of sitting down at the desk with Copernicus's manuscript, of turning its pages and, as a mathematician, realizing the enormous significance of what lay before him. We do know from Rheticus's writings that he was astonished by the sheer scope of Copernicus's work. In six lengthy sections, Copernicus had rebutted the entire Ptolemaic thesis and had demonstrated the superiority of the elegant heliocentric model—at least in his own view. But Copernicus was not yet ready to release his manuscript for publication; various elements had to be extended, planetary predictions improved, inconsistencies resolved before the work could be sent to the printer. And, as always, he was unnerved by the hostile reaction his theory might elicit. The idea had even occurred to Copernicus not to publish his work at all, but instead to make the manuscript available only to a few like-minded scholars such as Georg Rheticus.

At the time, Copernicus was sixty-six years old, and it must have occurred to Rheticus that if the three-decade-long project were not soon brought to a conclusion, it might remain forever incomplete. Rheticus made it his mission to coax the elderly astronomer into finishing the book. (The Yiddish word *nudge* comes to mind.) For the next two years, Rheticus rarely left the side of the man he referred to as *Domine Praeceptor*, "master teacher." Rheticus began to copy over the completed portions of the manuscript in the hope that his reluctant teacher might some-day consent to its publication. Had Rheticus not taken such an activist stance, Copernicus's work might have wound up in a dusty archive in Frombork, unpublished and unread. Indeed, near the end of his life, Rheticus is supposed to have remarked to a student of his: "You come to see me at the same age as I myself went to Copernicus. If I had not visited him, none of his works would have seen the light."

While Copernicus struggled with his revisions, he permitted his energetic young "student" to publish a summary of the massive work. Rheticus wrote feverishly, perhaps afraid that Copernicus might withdraw his consent, and in early 1540 the *Narratio prima,* or "First Report," of the Copernican theory appeared. To "grease the skids," as it were, Rheticus dismisses the old geocentric model through the words of twelfth-century philosopher Averroes: "The Ptolemaic astronomy is nothing, so far as existence is concerned; but it is convenient for computing the nonexistent." Then Rheticus rhapsodizes about the "harmony" of the heliocentric system and how Copernicus had managed to create a theory where the order and motions of all the celestial bodies are "linked most nobly together, as by a golden chain." Yet Rheticus also captures the frustration felt by both Copernicus and himself in trying to make sense of planetary motion: "The astronomer who studies the motions of the [planets] is surely like a blind man who, with only a staff [mathematics] to guide him, must make a great, endless, hazardous journey that winds through innumerable desolate places. . . . [He] will at some time, leaning upon it, cry out in despair to Heaven, Earth and all the

Gods to aid him in his misery." Rheticus's book sold briskly, a fact that might have convinced the ever-reluctant Copernicus that the criticism he feared might be muted. Apparently, scholars were eager to give his radical ideas a fair hearing.

In September 1541, while Michelangelo was putting the finishing touches on *The Last Judgment*, Copernicus entrusted Rheticus with a copy of his completed manuscript, which he had entitled *De Revolutionibus*, "On the Revolutions." Rheticus delivered the manuscript to a reputable printer in Nuremberg. He extended his leave of absence from the University of Wittenberg in order to proofread the book's printed pages and supervise production of the 142 woodblock illustrations.

In his preface, addressed to Pope Paul III, Copernicus confesses his longstanding fear of repudiation: "[T]he scorn which I had reason to fear on account of the novelty and unconventionality of my opinion almost induced me to abandon completely the work which I had undertaken." Yet later in the preface, as though having gathered some measure of strength from his own words, Copernicus turns combative: "Perhaps there will be babblers who claim to be judges of astronomy although completely ignorant of the subject and, badly distorting some passage of Scripture to their purpose, will dare to find fault with my undertaking and censure it. I disregard them even to the extent of despising their criticism as unfounded."

Midway through the printing of *De Revolutionibus*, Rheticus left Wittenberg for a more lucrative professorship in Leipzig. He enlisted the prominent Lutheran theologian Andreas Osiander to act in his stead. This same Osiander had previously corresponded with Copernicus, proffering the advice that if the astronomer was so fearful of ridicule, he should declare the contents of the book to be hypothetical, and not a literal rendering of the universe. Copernicus refused; evidently he believed his heliocentric creation was real. Nevertheless, the well-intentioned Osiander directed that his own anonymous apologia be inserted at the front of Copernicus's book. So every first-edition *De Revolutionibus* opens with the statement, ostensibly from Copernicus himself, that the heliocentric theory is purely hypothetical:

> [I]t is not necessary for the [book's] hypotheses to be true, nor even probable; it is sufficient if the calculations agree with the observations. . . . As far as hypotheses are concerned, let no one expect anything certain from astronomy, which cannot provide

it, lest he take as true something constructed for another purpose, and leave this discipline a greater fool than when he entered.

Rheticus was furious when he found Osiander's anonymous preface in Copernicus's book. Osiander had also fiddled with the title, which now read *De Revolutionibus Orbium Coelestium,* "On the Revolutions of the Heavenly Spheres," an allusion to the crystalline spheres in ancient cosmologies. Rheticus crossed out—in red ink!—both the illegitimate preface and the added title words in every copy he procured for colleagues.

Osiander's anonymous intrusion into Copernicus's work had its intended effect: Criticism of *De Revolutionibus* was generally subdued, and the book avoided censure by the Church until 1616. In that year, the work was placed on the *Index of Forbidden Books* when Galileo defended it as literal truth. Galileo and another fervent Copernican, Reverend Paolo Antonio Foscarini, were warned by Cardinal Bellarmine: "It seems to me that your Reverence and Signor Galileo would act prudently were you to content yourselves with speaking hypothetically and not absolutely, as I have always believed Copernicus spoke." Most astronomers at the time knew that Osiander was the true author of the anonymous preface. (Kepler found out through a handwritten annotation in a first-edition copy he had acquired.) Nevertheless, publishers into the nineteenth century continued to print the book with the unsigned preface.

De Revolutionibus did not take the literary market by storm. At the princely, modern-equivalent price of over $100, few could afford the 404-page book. As with Ptolemy's *Almagest,* only the mathematically adroit were able to wade through its calculational thickets. And what of the criticism that Copernicus had feared? Copernicus certainly did not become a religious martyr, as Victorian-era accounts implied; neither the Catholic clergy nor the Pope objected to his work. Lutheran fundamentalists were only moderately vocal. In the midst of other pressing matters, Martin Luther merely dismissed Copernicus as "that fool [who] wants to turn the whole art of astronomy upside down." As expected, opposing treatises appeared. Polymath Jean Bodin wrote in the late 1500s, "No one in his senses, or imbued with the slightest knowledge of physics, will ever think that the earth, heavy and unwieldy from its own weight and mass, staggers up and down around its own center and that of the sun." Even

Galileo, who in the early 1600s would become Copernicus's staunchest ally, initially supported the Ptolemaic system. Some of the more humorous critiques (at least, in retrospect) of Copernicus's science and mathematics are found in the hand-written annotations of surviving first editions: "Here Copernicus is dreaming!" "[Copernicus] deserves whips and lashes!" "The heavens would behave like a clown if they had to go as Copernicus wants!" There is only one documented instance of the willful destruction of a *De Revolutionibus*: Mediterranean pirates hurled overboard a first edition after they discovered that the captive ship's cargo consisted largely of books.

In May 1543, when a copy of the printed *De Revolutionibus* finally arrived in Frombork, the outcry that Copernicus had feared had become irrelevant. Copernicus had suffered a stroke and now lay on his deathbed. According to a letter Rheticus later received from Copernicus's longtime friend, Bishop Giese, Copernicus died only hours after his master work was placed before him. Whether Copernicus read Osiander's preface, Giese does not say.

In the center of all resides the Sun. For in this most beautiful temple, who would place this lamp in another or better place than that from which it can illuminate the whole at one and the same time? As a matter of fact, not inappropriately do some call it the lantern of the universe; others, its mind; and others still, its ruler. The Thrice-Great Hermes calls it a "visible god"; Sophocles's Electra, "that which gazes upon all things." And thus the Sun, as if seated on a kingly throne, governs the family of the planets.

The tone of the first volume of *De Revolutionibus*, as the above passage attests, is at times rapturous, as though in writing it Copernicus had been inspired by a divine truth. He describes the elements of his heliocentric plan as "so linked together that in no portion of it can anything be shifted without disrupting the remaining parts and the universe as a whole." "What a picture," he enthuses, "so simple, so clear, so beautiful." But in truth, Copernicus's model of the heavens turned out to be just as complicated—and in important respects, just as erroneous—as the old Ptolemaic system.

Book One of *De Revolutionibus* covers the basic tenets of the Copernican theory and why they represent an improvement over those of Ptolemy. Copernicus reassures his readers that the Earth can indeed be in motion without the dire consequences augured by the ancients. Rotation is a natural form of motion, Copernicus writes; the Earth's atmosphere and surface objects, including its human residents, all circle in unison with the body of the planet. It is our planet's spin that makes the stars *appear* to rise and set; like the central Sun, the celestial sphere is stationary. (Although Copernicus did not advance the idea, with the stars now at rest, there was no reason why they had to reside on a sphere at all, but could be scattered in space.)

Copernicus posits the Earth's orbital movement as the cause of many phenomena relating to the appearance of the planets, such as their varying brightness, retrograde loops, and occasional clusterings in the sky. He claims that the absence of such phenomena in the stars demonstrates their remoteness: "From Saturn, the highest of the planets, to the sphere of the fixed stars there is an additional gap of the largest size. . . . So vast, without any question, is the divine handiwork of the most excellent Almighty." It is here that Copernicus reignites the stellar parallax debate by echoing his predecessor Aristarchus's assertion that the distance to the stars is immeasurably great (although for unknown reasons Copernicus deleted the sole reference to Aristarchus's heliocentric model before he shipped off his manuscript to the printer). Copernicus erects a firm line of defense against Ptolemaic critics who would argue that the absence of stellar parallax precludes a heliocentric universe.

In Book Two, Copernicus describes the celestial coordinate system and presents a companion star catalogue. Book Three covers his revision of Ptolemy's estimate of the Earth's precession and what effect this has on the length and nonuniformity of the calendar year. (In Copernicus's time, calendar reform was a pressing issue; the running calendar date had fallen almost a week out of sync with the onset of the seasons, as determined by the position of the Sun.) In the remaining three sections of *De Revolutionibus,* Copernicus introduces his heliocentric accounting of lunar and planetary motions. He does away with the equant-point concept and the nonuniform planetary movement that Ptolemy had foisted upon the geocentric model. Instead Copernicus has each planet circle at constant speed in a small epicycle whose center moves uniformly around the Sun, a solution that had been proposed by Islamic astronomer Ibn as-Shatir two centuries earlier. (To bring his model into accord with the observations, Copernicus was forced to locate the center of each planet's orbit some distance away from the Sun. Thus, his "heliocentric" universe essentially wheeled around an empty spot in space—a dubious proposal at best.)

The Copernican system had a major advantage over its rival: The planets' order and spacing are no longer arbitrary. In Ptolemy's universe, there is no way to be certain whether Mercury or Venus is closer to the cosmic center. Also, the theorized size of the entire system is subjective: Each planet's deferent and epicycle supposedly lies within a hollowed-out crystalline sphere, but the gaps between successive spheres are unknown. (Ptolemy preferred tightly nested spheres with no extraneous space in between.) Copernicus's heliocentric model, on the other hand, mathematically places the known planets in their proper order and specifies the size of each planet's orbit. First comes Mercury at 0.38 times the Earth's distance from the Sun, followed by Venus at 0.72, then the Earth. Moving farther out, Mars falls at 1.5 times the Earth's distance from the Sun, then Jupiter at 5.2, and finally Saturn at 9.5.

Another indisputable triumph of the heliocentric model is its natural explanation of the retrograde motion of planets. The model arrays the planets in their modern order, based on how long it takes each planet to circle the Sun. The farther out the planet, the more slowly it moves in its orbit. Thus, Mars progresses more slowly around the Sun than does the Earth and displays a reverse loop in its movement across the sky each time it is "lapped" by our more rapidly orbiting planet.

Acceptance of Copernicus's heliocentric system was by no means immediate or even assured. With contrivances such as epicycles and eccentrically placed planetary orbits, the Copernican model's promised virtue of simplicity was rarely in evidence. And in the end, his planetary predictions were little better than Ptolemy's. Having donned the Aristotelian straitjacket of uniform circular motion, Copernicus was unable to create a truly modern heliocentric universe. In the rigidity of his preconceptions, Copernicus is more readily identified with his forebears of antiquity than with his successors Galileo, Kepler, and Newton. Overall, we can appreciate the aesthetic appeal of the heliocentric arrangement, but we are left with a nagging sense that Copernicus's rendition is flawed. *De Revolutionibus* may have triggered the so-called Copernican Revolution, but it was not the revolution itself.

It is hard to say precisely when the Sun-centered arrangement of the universe came to be regarded as fact. Indeed, acceptance of any controversial theory spreads gradually as scientists make confirming observations and theoretical refinements. Tycho Brahe, the foremost observational astronomer of the 1500s, could not abide a heliocentric universe and offered his own geocentric plan in competition with Copernicus's model. Galileo became a fervent believer in the early 1600s, following his telescopic exploration of the solar system. His contemporary, Johannes Kepler, also preferred the heliocentric arrangement, although his studies of the movements of Mars led him to discard Copernicus's cherished assumption of uniform circular motion. Kepler's introduction of *elliptical* planetary orbits finally broke the ancient Aristotelian stranglehold on cosmic research and gave us the heliocentric model we know today. Kepler's recognition that the Sun, besides being central, also exerted a physical influence on the planets paved the way for Isaac Newton's contributions to the subject. In the late 1600s, with Newton's theories about motion and gravity (subsequently elevated to "laws," the highest accolade science can bestow upon explanations of natural phenomena), the heliocentric universe at last gained its physical foundation: The massive Sun held the planets in its gravitational grip and determined the essential aspects of their orbital motion.

In Newton's day, the Ptolemaic system and the Keplerian version of the Copernican system were taught side by side in the universities of the world. But the pendulum of belief had swung irreversibly to the Copernican side. In the minds of most scientists, the heliocentric universe had become fact. The Earth was free at last to sail unhindered around the central Sun, just as Aristarchus had envisioned it almost two thousand years earlier. Yet there remained a crucial missing element in what was otherwise a complete and compelling picture of the universe: Not one shred of indisputable *observational* proof existed that the Earth moved through space. Here then was the holy grail of many an astronomer. To prove that the Earth in fact revolved in a wide orbit around the Sun, the parallax of just one star—any star—had to be detected. The hunt for stellar parallax was on.

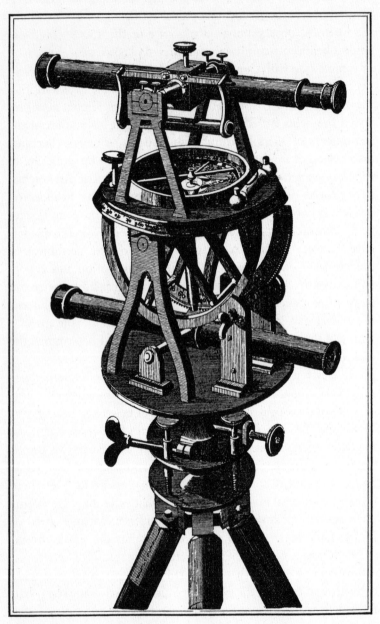

A theodolite by instrument maker Jesse Ramsden. From Repsold (1908), based on a 1791 engraving.
Source: *Wolbach Library, Harvard University.*

4

Crossed Eyes and Wobbling Stars

The aspects of things that are most important for us are hidden because of their simplicity and familiarity. One is unable to notice something because it is always before one's eyes.

—*Ludwig Wittgenstein*

Many things are known, as some are seen, that is by Parallaxis, or at some distance from their true and proper beings.

—*Sir Thomas Browne,* Christian Morals, *1682*

★★★

P
A
R
A
L
L
A
X

Close your right eye. Hold a finger upright a couple of inches in front of your nose such that your finger appears somewhere to the right of the word **PARALLAX** above. Focus on the word, not on your finger. Now close your left eye and open your right. Your finger appears to the left of the word **PARALLAX.** If you alternately open and close each of your eyes, your finger will seem to shift repeatedly from left to right and back again. This is the phenomenon of parallax. It's the same phenomenon that lets us cheat when weighing ourselves; by shifting our head to the right, the indicator on the bathroom scale appears superimposed against the smaller numbers to the left. (Digital-readout scales aren't fooled by such trickery.) Parallax occurs because each eye provides a unique vantage point on the world and thus sees the world a little differently. So why is it that when we view our surroundings with both eyes open, we don't see a confusion of doubled-up images?

Light entering the pupil of the eye is focused by a flexible lens toward the back of the eye. Here lies the retina, a thin layer of light-sensitive cells that serves as the "screen" onto which images of the outside world are projected. The retina chemically converts light energy into electrical impulses, which race to the brain's visual cortex via the optic nerve. Given the spacing between our eyes and the attendant parallax this imposes on objects being viewed, the retinal images of any given scene will not be quite the same in both eyes. How does your brain contend with these two conflicting takes on the real world?

Starting from infancy, the visual cortex of your brain has trained itself through sensory experience to combine the dual retinal images into a single coherent, three-dimensional "picture." It's easy to uncover the inherent duality of this amalgamated picture. Look at the page with both eyes open. Now gently press the outer part of one eyelid with your finger. The writing on the page doubles up. Fed this unexpected input, your brain doesn't know how to assemble the two retinal images—one horizontal, one canted— into a coherent whole. (Were you to maintain the pressure against your eyelid, your brain would eventually learn how to properly integrate the double images.) When I first discovered this trick as a kid, I wondered which of the two worlds was the "real" one. If you chose the horizontal image as real and the canted image as distorted, try this: Repeat the experiment, but close your unpressed eye; the canted world appears quite normal, doesn't it?

Without the early lessons of interlinked stimuli—visual, tactile, and auditory—our surroundings would appear to us as a flat jumble of colors, shapes, and movement, devoid of a third dimension. An infant swatting at an out-of-reach mobile suspended above the crib has not yet acquired depth perception. In certain situations, adults lack the ability to perceive distance. Neurologist Oliver Sacks writes movingly of Virgil, fifty years old and blind since infancy, whose vision was restored through surgery:

> *Virgil told me later that in his first moment he had no idea what he was seeing. There was light, there was movement,*

there was color, all mixed up, all meaningless, a blur. Then out
of the blur came a voice that said, "Well?" Then, and only
then, he said, did he finally realize that this chaos of light and
shadow was a face—and, indeed, the face of his surgeon. . . .
[Virgil] was able to see but not to decipher what he was seeing.

Virgil also had no conception of distance; he would recoil at
the sight of a bird because he couldn't be sure how far away it
was. Lifelong rainforest dwellers are not accustomed to seeing
very far; the dense vegetation of their environment limits their
line of sight to a few feet. When confronted with an unobstructed vista, they will reach out and try to touch mountains on the
horizon. Sacks concludes that we "are not given the world; we
make our world through incessant experience, categorization,
memory, reconnection."

Essayist Annie Dillard writes that for infants or the newly
sighted, "vision is pure sensation unencumbered by meaning."
She imagines her own transition into three-dimensional reality:
"I'm told I reached for the moon; many babies do. But the color-
patches of infancy swelled as meaning filled them; they arrayed
themselves in solemn ranks down distance which unrolled and
stretched before me like a plain. The moon rocketed away. I live
now in a world of shadows that shape and distance color, a world
where space makes a kind of terrible sense."

In his memoirs, nineteenth-century physicist Hermann von
Helmholtz recounts an experience he had when he was only two
years old. He was strolling in a park with his mother and saw what
appeared to be a toy turret with tiny human figurines peering out
over the railing at the top. He asked his mother to give him one of
the figurines to hold. She explained that the turret was not a toy,
but a real turret about a mile away, with real people on top. In a
flash, Helmholtz recalled, the third dimension unfurled before his
eyes, and his view of the world was forever altered. He had learned
to perceive depth.

When it comes to distance perception, even an experienced
observer can be fooled. This was driven home to me a few years
ago while I was cruising along the interstate. I saw what looked
like an enormous hot-air balloon off in the distance. The balloon
was so far away that I couldn't even make out the passenger bas-
ket I believed to be suspended below it. The huge sphere looked
positively serene, floating lazily in the sky, and I imagined what it
would be like to see the world from up there. But as I approached

the balloon, I could tell that something was wrong. My visual sensations of the balloon didn't seem right. The balloon was rapidly picking up speed, as though an invisible hand were accelerating it sideways. Soon it appeared to be moving impossibly fast across the sky. What in the world was going on? Then, in an instant, my perception of the scene changed dramatically. This was no giant, passenger-carrying balloon I was looking at; it was a room-sized advertising balloon. And there, as I drove past, was the rope tethering it to the ground. The balloon's perceived size, distance, and motion were an optical illusion, a complete fiction impressed upon it by my own brain.

D
I
S
T
A
N
C
E

This next experiment reveals an essential feature of the parallax phenomenon, one that is at the heart of this book: the ability to use parallax to measure distance. Raise your finger again to your nose and sight the word **DISTANCE** above. Alternately blink your eyes as before and observe the apparent shift of your finger between left and right. Note the magnitude of the shift. Now move your finger a little farther away and blink again. The parallax shift is smaller than it was before. Move your finger farther yet, and the parallax shift becomes even smaller.

The results of this experiment suggest the following rule: *The farther an object, the smaller its parallax.* When estimating an object's parallax, it's often helpful to have something in the background against which the object's parallactic shift can be gauged. Here, that purpose is served by the word **DISTANCE** on the page (although it, too, displays its own parallax). For a tree, the background object might be a distant church steeple or a much more remote tree. For a heavenly body such as the Moon or a "nearby" star, the field of faraway stars forms the background.

The converse of this rule is also true: *The smaller the parallax, the farther the object.* Here is the key to measuring the distances of objects around us, from a flower a few inches away to a star trillions of miles in outer space. *The smaller the parallax, the farther the object.* Clearly, for us to perceive distance most effectively, our brain must be able to gauge the magnitude of the parallax sensed by the eyes. What is the physiological basis behind this process?

Two muscle groups work the magic. The ciliary muscles stretch or contract to focus the eye lens by altering its shape: thicker for nearby objects, thinner for distant objects. This phenomenon is known as *accommodation*. The rectus and oblique muscle groups—the ones directly responsible for parallax sensing—rotate the eyeballs, allowing them to converge on an object. You can feel the rectus and oblique muscles at work by pressing lightly against your eyelid while sweeping your eyes in various directions. Better yet, videotape yourself while sighting objects at a variety of distances. For close-up objects, the videotape should reveal a marked convergence of your eyes, that is, you'll appear cross-eyed. (Looking in a mirror won't work because you can't focus on an object while simultaneously observing your reflection.) To effectively perceive distance and create a three-dimensional, or stereoscopic, view of your surroundings, your brain senses the convergence of your eyeballs, then considers a host of additional cues, including accommodation, movement of the object, shadowing and perspective, binocular disparity (the different appearances of the two retinal images), and the brain's own catalogue of experience.

Every creature that has two front-facing eyes or two eyes on the same side of its head—primates, owls, flounders—has the potential to detect parallax. Even animals whose eyes fall on opposite sides of the head—parakeets and certain reptiles—can rotate their heads for this purpose. The evolution of stereoscopic vision can be linked to the survival and advancement of our species, from grasping for food to hunting prey to evading predators. If you were a caveman hurling a spear at a wild boar, it would be helpful to be able to judge the boar's distance. This "distance-feeling," as physicist George Gamow has termed it, assists us in more modern-day pursuits: threading a needle; driving; playing sports such as tennis, baseball, badminton, or basketball; and even reaching for someone's hand. If there is a disadvantage to be found in the parallax-sensing component of the human visual system, it is the limited range over which it is accurate, some tens of feet. But there are ways to extend this range—ultimately, all the way out to the stars.

BROADENING THE BASELINE (I):
RANGE FINDERS

The eyes of a human being are typically separated by about two and a half inches. This is your natural-born baseline that governs—and limits—your perception of distance in the world. Beyond about a hundred feet, an object's distance overwhelms the "capacity" of your rather close-set baseline. It's a simple matter of geometry: With ever more distant objects, the convergence of your eyes becomes too minuscule to convey useful parallax sensations to the brain. Your eyes are virtually parallel whether focused on an object a few hundred feet away or a mile away; your brain is forced to rely on alternate stimuli and on past experience to determine the object's distance. The range of our ocular parallax-sensing mechanism could be extended if it were possible to widen the eye-to-eye baseline. Impossible? Of course. But let's not be so literal. We can *effectively* increase the separation of the eyes by the magician's long-time mainstay: mirrors.

My father owned an IDEAL-brand binocular range finder made by a long-defunct company on Long Island. He used it to determine where to set the focus of his camera. (Range-finder cameras didn't exist in those days—at least not in our household.) The IDEAL range finder was small enough to fit in my shirt pocket and became one of my favorite "tech-toys" when I was a kid, on par with my Erector set and crystal radio. Magically, it seemed, this device could sense the distance to any object at which it was pointed.

My father's range finder now belongs to me. It has a rectangular plastic body, all black, about the size of a pack of Life Savers. At the far ends of one of the long sides is a pair of little circular apertures through which light enters. On the opposite side is a viewing hole and a numbered dial calibrated from 2.5 to 100 feet. Peering at an object through the range finder, you see two images of the object standing side by side. Turn the dial and the images gradually merge; the dial now indicates the object's distance.

As a kid, I wondered how this little device managed such a feat. So like any good scientist, I took it apart. (My father was an engineer and was always salvaging things to see what was inside. Once, when I was in high school, I removed the motor from our vacuum cleaner and used it to suck the air out of a big mayonnaise jar. It was part of a science project about the effect of a low-pressure, "Martian" atmosphere on the growth of plants. The

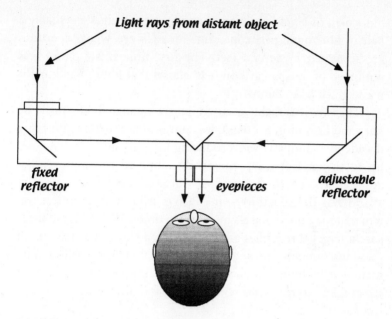

Principle of the binocular range finder.

plants didn't survive.) The workings of the range finder are shown in the illustration above. Behind each aperture is an angled mirror that reflects light toward the center of the range finder, where it is further reflected into the eye. Thus, the eye sees images from both apertures simultaneously. One of the angled mirrors is fixed; the other is glued to a flexible metal strip that bends increasingly when the dial is turned, angling in the mirror. Taken together, the mirrors form a mechanical version of a pair of converging eyes.

A much expanded version of my "toy" range finder came into military use in the late 1800s. Prior to that time, artillery gunners would shoot "short," then converge upon the target with repeated rounds. ("Acoustic ranging" was also used, where the operator timed the delay between an enemy gun's visible flash or puff of smoke and the sound of the gun's report; since light travels much faster than sound, the greater the delay, the farther the gun. The acoustic method never worked all that well; once the battle was engaged, it was impossible to tell in the thunderous din which sound belonged to which flash or smoke puff. Once smokeless powder was introduced, acoustic ranging was abandoned.) With both the range of cannons and the cost of projectiles escalating, a more accurate targeting method was needed. The solution: the binocular, or single-observer, range finder, which is essentially a pair of horizontal periscopes, one for each eye, whose apertures are separated by several feet. The first time I saw one, I was reminded of outsize cartoon field glasses that Elmer Fudd might use to track Bugs Bunny.

The first reasonably accurate portable range finder, made by Barr and Stroud in Scotland, was put into use in 1888. Within a decade, the company was producing instruments with baselines up to fifteen feet for permanently mounted fortress or naval applications. The U.S. Army M7 portable range finder, used in World War II, had a thirty-nine-inch baseline and an optical system that magnified objects by fourteen times. The M7 could measure a target three miles away to within a tenth of a mile. In all these instruments, the angle of an internal mirror is adjusted by turning a calibrated knob until the split images viewed through the eyepiece coincide; the object's range is indicated on a mechanical counter.

In principle, a binocular range finder can be made arbitrarily large; in practice, it's not so easy. If the apertures are spaced many feet apart, the optical system can easily fall out of alignment, especially when the ungainly instrument is lugged from place to place. The range finder is fundamentally limited by its design: Images from two widely spaced apertures have to be viewed simultaneously. This severely restricts the maximum effective baseline of the device, and hence its maximum range. But what if we didn't have to see the object with both eyes at the same time? What if we could somehow record an object's position as seen from one vantage point, then simply move to another vantage point and look again? Such is the essence of surveying.

BROADENING THE BASELINE (II): SURVEYING

As official surveyor of Culpeper County, Virginia, George Washington certainly knew the parallax mantra: *The smaller the parallax, the farther the object.* Surveyors employ parallax when trying to determine the distance to a remote object. They call the process *triangulation,* but it's essentially the converging-eye principle all over again, just on a larger scale. Unlike stereoscopic vision or binocular range finders, surveying methods do not require *simultaneous* observations from both ends of a baseline. Instead a sturdy spotting scope, called a theodolite (pictured at

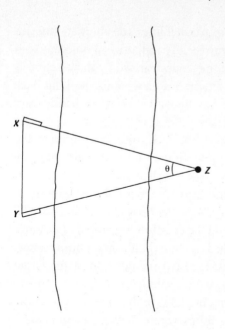

Parallax of a tree.

the beginning of this chapter), is employed to measure an object's position, first as seen from one end of the baseline, then from the other. With such an instrument, angles can be measured to within one arc-second, or $1/3{,}600$ of a degree, so a half-inch sideways shift of an object can be detected from two miles away.

Here's how we might measure the width of a river by sighting a tree on the opposite bank. First we establish a baseline and measure its length; as pictured here, the baseline runs between points X and Y, approximately parallel to the riverbank. Together with the tree, at point Z, the three points form a triangle. We set up our theodolite at point X and point the instrument at the tree. Using the scale on the theodolite, we measure the angle YXZ between the baseline and the tree. Then we move the instrument to point Y and measure the angle XYZ. (In this instance, the triangle is symmetrical and the angles YXZ and XYZ are the same, but that's not always the case.) Now we know the length of one side of the triangle—our baseline—plus the angles at which the other sides branch off from it. At this point, we could construct a scale diagram to determine the lengths of the two unknown sides, and thereby get an estimate of the river's width. (Or we could use the rules of trigonometry to compute the exact, "straight-across" width.)

The angle labeled θ is the difference in the tree's apparent position when seen from point Y and when seen from point X. It is the tree's parallax as seen from the ends of the baseline. Let's switch our vantage point to the other side of the river so that we are looking back at the baseline across the water. Angle XZY spanned by the baseline, as seen from our new viewpoint at the tree, is equal to the parallax angle θ. (The relevant geometric rule is: If a line intersects a pair of parallel lines, the alternate interior angles are equal.) In other words, the parallax of an object as seen from the baseline is equal to the angle spanned by the baseline as seen from the object. This alternative viewpoint—looking back from the object toward the baseline—will come in handy later when we discuss stellar parallax.

Shortly after World War I, the U.S. military combined the essentials of the binocular range finder and surveying into one system and applied it to a strategic problem: how to target coastal artillery guns on enemy battleships. The method utilized a ship's parallax to reveal its distance. A telescope was placed at each of two observation stations along the coastline, separated in distance by one-quarter to one-third of the gun's maximum range. Observers there simultaneously noted the position of the target ship, as seen from their individual vantage points. Every thirty seconds (since a ship is a moving target), they relayed the information to a central plotting room, where the lines of sight were drawn on a map and targeting instructions were issued. The sixteen-inch guns at these installations could sink a battleship with a single, well-targeted round. But they never had to. The psychological deterrent value of the guns and their parallax-based targeting system was sufficient to keep enemy ships at bay.

Conventional surveying techniques have also been used extensively in mapping the globe. Triangulation has even been applied above the Earth's surface to meteors streaking through the atmosphere. The fiery meteor is photographed from a pair of stations many miles apart; because of parallax, each photo will show the meteor in a different location among the stars. From the magnitude of the parallax shift, we are able to deduce the meteor's height above the ground.

From the atmosphere, it's only a short jump to the ultimate realm of parallax studies: outer space. So let's scale up our baseline once again and head away from the Earth.

BROADENING THE BASELINE (III): SOLAR-SYSTEM PARALLAXES

Have you ever glanced out your car window while driving, only to see the Moon keeping up with you? How can this be? Lamp posts, pedestrians, houses, trees—all appear to sweep past us as we drive down the road. But not the Moon. Even when you're speeding along at sixty miles an hour, where every passing minute shifts your vantage point by another mile, the Moon still looms in the distance, always in the same position.

The Moon's strange behavior has to do with parallax—or, actually, the absence of parallax. The Moon is so far away that no parallax shift is evident to the naked eye over a baseline of only a few miles; its position always appears the same. (The Moon does in fact move across the sky as the Earth turns, but that has nothing to do with parallax.) The first successful measurement of the Moon's parallax was carried out in the second century B.C. by the great observational astronomer Hipparchus, whom we've already met. Hipparchus accomplished this feat in a particularly clever way. The solar eclipse of March 14, 189 B.C., was visible over a wide swath of the Mediterranean region. Witnesses living near the Hellespont, the narrow strait off northwestern Turkey, reported that the eclipse was total; that is, the Moon totally covered the Sun's face. However, observers in Alexandria saw only four-fifths of the Sun obscured by the lunar disk. In this straightforward account, Hipparchus saw the opportunity to determine the parallax—and thus the distance—of the Moon. He assumed that the Sun is sufficiently distant that it presented no measurable parallax to the eclipse watchers. In other words, during the several minutes of maximum eclipse, the Sun served, in effect, as a stationary marker against which the Moon's parallax could be gauged; any difference in the eclipse's appearance from the two terrestrial vantage points must have stemmed from the parallax of the Moon alone. Simply put, the Alexandrians saw the Moon in a slightly different position than did their counterparts on the shores of the Hellespont.

One-fifth of the Sun had remained visible in Alexandria; thus, the Moon's position as seen from there had been shifted by an amount equal to one-fifth of the Sun's disk, about a tenth of a degree. Therefore, one-tenth of a degree is the Moon's parallax over a baseline extending between the Hellespont and Alexandria. Combining this parallax with the latitudes of the Hellespont and Alexandria and also with the reported altitude of the Moon at each site, Hipparchus deduced the Moon's distance: between thirty-five and forty-one Earth-diameters. The true value is approximately thirty Earth-diameters. Respectably close, considering that the work was carried out more than 2,000 years ago.

The Moon's parallax can also be measured by conventional triangulation: two observers sighting the Moon simultaneously from two widely separated places on the Earth. The first truly long-baseline determination of the lunar parallax was carried out in 1751 by French astronomer Nicolas Louis de Lacaille at the Cape of Good Hope, in cooperation with astronomers back in Europe. Lacaille found that the Moon displays a parallax of approximately 2 degrees to observers situated on opposite ends of the Earth, that is, over a baseline of 8,000 miles. Two degrees is equivalent to a sideways shift at arm's length of about an inch, quite easily discerned.

Looking deeper into space, astronomers had long sought parallaxes of planets. And for just as long, observers had deluded themselves into believing they had detected them. The parallaxes of even the nearest planets—Mercury, Venus, and Mars—are only a fraction of the Moon's parallax because these worlds are much more remote. Venus, for instance, which swings closer to Earth than does any other planet, is never less than about one hundred times the lunar distance; hence, its parallax will be at most one-hundredth of the Moon's. Nevertheless, the motivation to measure a planet's parallax is compelling. In the Copernican solar system, the *relative* sizes of all the planetary orbits are uniquely specified. For example, the orbit of Venus is about three-quarters as wide as the Earth's orbit, that of Mars about one-and-a-half times as wide, and Jupiter's about five times as wide. If the *actual* size of just one planet's orbit can be measured, then the actual sizes of all the others automatically follow. As a result, the true dimensions of the Copernican planetary system become known, as does the most fundamental solar system measure of all: the distance from the Earth to the Sun, designated the *astronomical unit*. All it takes is a reliable parallax measurement of one planet. Any planet.

Mars was the earliest target of serious planetary parallax investigation. The best time to measure the parallax of Mars is during a planetary configuration known as *opposition*, when the Earth passes directly between Mars and the Sun. At opposition, Mars swings as close to our planet as it ever does, and its parallax rises to a maximum. The great Danish observer Tycho Brahe, who will figure prominently in the *stellar* parallax story, attempted the parallax of Mars during the predicted opposition of 1582. Tycho noted Mars's position among the stars twice daily: once before dawn and again after sunset, the Earth's rotation having swept him in the interim along a planetwide baseline. Although he claimed to have achieved success, Tycho repeatedly altered his parallax numbers over the years. We know today that Tycho's purported measurement was spurious; the true parallax of Mars—an apparent shift of only 0.012 degrees when viewed over a baseline as wide as the Earth—is too small to have been detected by Tycho's instruments.

Almost a century later, in 1672, Giovanni Domenico Cassini, director of the Royal Observatory in Paris, launched an effort to measure the Martian parallax during that year's opposition. While Cassini recorded Mars's position from Paris, his assistant, Jean Richer, did the same from the island of Cayenne along the South American coast, in what is now French Guiana. The two astronomers concluded that Mars was about 4,000 Earth-diameters away at opposition and that the Sun-Earth distance—the astronomical unit—was approximately 87 million miles, respectably close to the modern-day value of 93 million miles. The near agreement, however, is coincidental; uncertainties in Cassini's and Richer's position measurements were so large as to render the result inconclusive. In scientific research, it isn't enough to get the right answer; the researcher has to convince a generally skeptical community of colleagues that the answer is the result of a valid and well-performed methodology. Even then, unknown sources of error might lurk within the raw observations, errors that sometimes lie hidden for generations. Especially in the past, when the scientific method was in its formative stages, a researcher's reputation inevitably lent weight to the experimental conclusions. Trustworthy results were assumed to be the norm for scientists with established "track records"—sometimes rightly so, and sometimes not. In the late 1600s, Cassini was deservedly among the most respected astronomers of the age. His 87-million-mile astronomical unit was generally accepted as correct well into the eighteenth century, even though the number was, in the words of historian Albert van Helden, nothing more than "a convenient estimate wrapped in the cloak of authority."

One astronomer who did reject Cassini's parallax of Mars—and by implication Cassini's derived distance to the Sun—was England's second Astronomer Royal, Edmond Halley. Halley proposed an alternative means of calibrating the astronomical unit: measuring the parallax of Venus. Venus swings closer to the Earth than does Mars, yet it is a more challenging parallax target because of its proximity to the Sun. During opposition, Mars always stands in the *night* sky, since by definition it is diametrically opposite the Sun in this configuration; therefore, stars provide the fixed grid against which Mars's position can be measured. By contrast, when an inner planet such as Venus passes closest to us, it always lies directly *between* the Sun and the Earth; from our viewpoint, Venus will appear during the light of day. How then can its parallax be measured when there is no starry grid against which to gauge its position?

The solution to this dilemma arose, strangely enough, from a case of youthful exuberance on Edmond Halley's part. In November 1676, when he was only nineteen years old and a student at Oxford, Halley traveled to the remote south Atlantic island of St. Helena, expressly to view parts of the sky inaccessible from England. Situated 1,200 miles west of Angola, this tiny volcanic outcropping served as a way station for vessels of the East India Company and, from 1816 until 1821, as Napoleon's final place of exile. "One can speak lightly of going to St. Helena," writes astronomer David S. Evans, "but one can only wonder at the skill or luck of the sailors with only the crudest of navigational techniques in ever finding the place."

Halley set up his equipment on the rugged slopes of Diana's Peak, the island's 2,700-foot summit. Among Halley's many observations during his eighteen-month-long stay on St. Helena was a predicted *transit* of the planet Mercury across the Sun's face on November 7, 1677. Passing directly between the Earth and the Sun, Mercury's tiny, silhouetted disk could easily be seen through a telescope against the Sun's glowing surface. (When viewing the Sun with a telescope, astronomers either projected the blinding image onto a white screen or fitted the telescope with a darkening filter.) Astronomer James Gregory had first proposed in 1663 that the circular outline of the Sun might provide a reference mark against which Mercury's parallax might be measured. Viewing the transit from St. Helena, Halley envisioned a network of astronomers observing a transit simultaneously from widespread locations on Earth. Each astronomer would record the exact time Mercury strayed within the solar outline, a time that would differ from observer to observer because of their divergent vantage points. The process would be repeated when Mercury exited the solar disk. Pooling their timing measurements, astronomers could accurately deduce Mercury's parallax. When Halley returned to England in 1678, he found the need for coordinated effort starkly evident: Only one other astronomer in all of Europe had observed the transit besides himself.

Halley's cooperative observing scheme was even more promising for Venus. Because of its proximity to the Earth, Halley pointed out, Venus's silhouetted disk would appear much

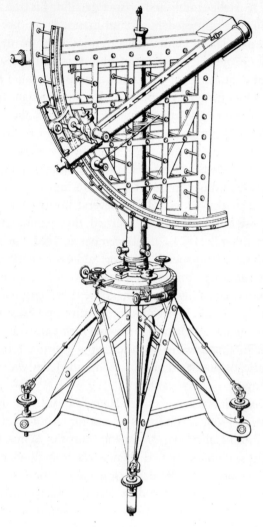

Portable quadrant from the early 1800s. From Pearson (1824).
Source: *Wolbach Library, Harvard University.*

larger than Mercury's. Also, the observed ingress or egress time of Venus across the solar outline should differ from observer to observer by as much as five minutes, so timing errors of several seconds could be tolerated. Equipment needs were minimal: an accurate clock and a portable position-measuring telescope, called a quadrant. The resultant parallax of Venus, Halley predicted, would be accurate to within 1 percent and would establish the size of the solar system with a surety that Cassini's Mars observations could not.

In 1716, Halley published a paper detailing his transit method and urging an unprecedented, multinational venture to determine the parallax of Venus. Advance planning was necessary to ensure the proper deployment of observers and the subsequent sharing of data. Personal and political rivalries would have to be set aside. Only through cooperation would the scientific mission succeed. Transits of Venus occur rarely: in pairs, about once per century. (Venus's orbit is inclined relative to our Earth's, so transits take place only during those sporadic instances when Venus slips directly between the Earth and the Sun.) Halley predicted that Venus would next transit the Sun on June 6, 1761, and June 3, 1769. Each transit would last several hours. If astronomers missed these opportunities, he warned, they would have to wait until December 9, 1874, and December 6, 1882, for their next chance. In his 1716 paper, Halley was addressing a future generation of astronomers, for he knew that neither he nor many of his counterparts would live to witness the 1761 passage of Venus across the Sun's face. Nor would he be alive on Christmas Eve in 1758 to witness the return of the comet now named in his honor, the comet he first saw in 1682 and whose eventual return he had mathematically predicted in 1705. Edmond Halley was a forward-thinking man.

In 1761, nineteen years after Halley's death, and again eight years later, in 1769, astronomers set out for remote corners of the Earth to fulfill Halley's vision for the transit of Venus. Observers from eight nations stationed themselves from England to Baja California; Norway to Peru; Hudson Bay to the Cape of Good Hope; Siberia to the East Indies. Nevil Maskelyne, Astronomer Royal in 1769, embarked for Halley's inspirational haunt, the island of St. Helena. Charles Mason and Jeremiah Dixon, who between transits surveyed their famous line through the Pennsylvania countryside, traveled to South Africa. Captain James Cook sailed his *Endeavor* across the Pacific to deposit astronomer Charles Green among the Tahitian palms. Misfortune dogged French astronomer Guillaume Le Gentil, who missed the 1761 transit when his observing site, the Indian settlement of Pondicherry, fell to the British. Le Gentil chose to settle in India rather than risk missing the 1769 transit. Imagine his grief when a curtain of clouds rolled in on the appointed day. Le Gentil sulked back to his native France only to find that he had been declared dead, his assets distributed among his relatives, and his post filled by another.

In all, 150 observations of Venus's transits were recorded worldwide. Contrary to Halley's hopes, the results did not prove definitive. The turbulence of Earth's atmosphere and various optical distortions made it hard for observers to judge the precise moment Venus's silhouette entered the solar disk. Nevertheless, from Venus's somewhat imprecise parallax, the Sun was determined to lie about 91 million miles from Earth, in good agreement with both Cassini's earlier estimate and the currently accepted value for the astronomical unit.

The next transit of Venus is scheduled for June 8, 2004, and will be best viewed from Eastern Europe and Western Asia. Following the transit of 1882, William Harkness of the U.S. Naval Observatory peered ahead to the one that awaits us now. "There will be no other [transit of Venus]," Harkness wrote, "till the twenty-first century of our era has dawned upon the earth, and the June flowers are blooming in 2004. When the last transit occurred the intellectual world was awakening from the slumber of ages, and that wondrous scientific activity which has led to our present advanced knowledge was just beginning. What will be the state of science when the next transit season arrives God only knows." The transits of Venus are the ticking of a celestial clockwork that beats century-long seconds in the life of our civilization.

The only other solar system bodies that stray within triangulation range of the Earth are a few asteroids. Most notable of these is Eros, named for the god of love in Greek mythology. Eros is an irregular, twenty-one-mile-wide chunk of rock whose orbit around the Sun occasionally carries it as close as 14 million miles from our planet. That's nearly sixty times the Moon's distance, but still sufficiently near to display a measurable parallax to widely spaced observers on the Earth. As with the planets, a determination of an asteroid's parallax permits the calculation of the astronomical unit.

Here's the essence of the procedure. The orbital characteristics of both the Earth and Eros are well known: size, shape, orientation in space, and period (the time it takes to swing entirely around the Sun). On average, Eros is about 1.5 times farther from the Sun than is the Earth, although its looping orbit at times brings it closer to the Sun than the Earth ever gets. Eros's orbital period is 642 days, compared to the Earth's 365 days. Astronomers know precisely where Eros and the Earth are located in their respective paths at all times. In particular, using the mathematical rules of planetary motion, they can predict the separations—in *astronomical units*—between the two bodies at any chosen moment. Therefore, a simultaneous measurement of the number of *miles* separating Eros and the Earth would reveal how many miles comprise the astronomical unit. In 1931, an international team of astronomers set out to do just that. During that year, as Eros swung "close" to the Earth, observers from fifteen countries worked in concert to measure its parallax. Nearly 3,000 photographs were taken of the moving asteroid as it crept across the starry background. Another ten years was spent analyzing the images. The result: at closest approach, Eros was about 16 million miles from the Earth. From this, the astronomical unit worked out to be nearly 93 million miles long. Suppose the Sun were represented by an eleven-foot-wide weather balloon; to scale, our Earth would be a Ping-Pong ball orbiting a quarter-mile away. On this same scale, Eros in 1931 would lie three-quarters of a football field away from the Ping-Pong ball and would be too small to be seen by the unaided eye.

NASA recently succeeded—appropriately on Valentine's Day, 2000—in slipping its Near Earth Asteroid Rendezvous spacecraft into orbit around Eros. Mission operators kept track of the spacecraft's orbital position the "old-fashioned" way: They triangulated on Eros's surface features. Just before Valentine's Day, 2001, the ship settled gently onto the asteroid's boulder-strewn landscape.

The widest possible baseline that can be achieved on the Earth is 8,000 miles—one Earth-diameter. If only there were a way to expand this worldwide baseline. . . . In the geocentric model of the universe, we have reached the limit of parallax exploration; whichever parallaxes are too small to be measured, given our stationary planet's baseline, will never be measured. But if the heliocentric arrangement of the universe is true, the Earth glides around the Sun in a huge orbit. Our planet's motion carries us to vantage points in space that are inaccessible to inhabitants of a stationary world. With a moving Earth, we can stretch our 8,000-mile baseline more than *twenty-thousandfold* and project our parallax-detecting powers into the realm of the stars.

BROADENING THE BASELINE (IV):
STELLAR PARALLAX

Suppose Aristarchus is right: The Earth circles the Sun. The width of the Earth's orbit is twice its 93-million-mile radius, or 186

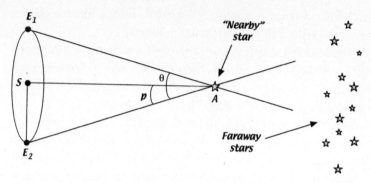

Parallax of a star.

million miles. Wherever the Earth is presently situated in its orbit, half a year from now it will be diametrically opposite, 186 million miles from its former position. Over a six-month time span, a single observer on the Earth gains vantage points on the heavens that are as much as 186 million miles apart. Therein lies the impetus for astronomy's epic struggle to observe the parallax of a star. If a surveyor can triangulate a tree over a baseline of a hundred feet, surely an astronomer can triangulate a star over a baseline of 186 million miles.

Pictured here is a representation of the Earth in its orbit around the Sun, labeled S. (The nearly circular orbit appears foreshortened in the diagram.) At a given time, the Earth is situated at point E_1; six months later, it has swung to the opposite side of its orbit at E_2. The distance between the Earth and the Sun, SE_1 or SE_2, is one astronomical unit: 93 million miles. Far off in space is a star, labeled A. When the Earth is at E_1, we point our telescope toward the star and note where it appears relative to other stars (presumably) behind it. We wait six months, then reobserve star A from our new vantage point at E_2. If the star is sufficiently close, its position will have shifted relative to the background stars. (If these background stars are very far, they will have no discernible shift in position. Recall: The farther the object, the smaller the parallax.)

A telescope that had been directed at star A from vantage point E_1 is tilted differently when sighting the same star from E_2. The star's position appears to have changed from one vantage point to the other. This difference in position is the star's parallax, denoted by the angle θ. However, astronomers have agreed to define a star's parallax formally as *half* of the angle θ, that is, half of the star's total observed shift in the sky. Let's label this "official" astronomical parallax angle **p**. (Bear with me a while longer, for there *is* a method to this particular madness.)

The reason for defining stellar parallax as half the star's total shift is revealed by considering the view *from* star **A**. Looking back toward the solar system, we would see the Sun with the Earth revolving about it. The angle spanned by the astronomical unit, as seen from the star, is equivalent to the star's parallax angle **p**, as seen from the Earth. Effectively, the astronomer's parallax definition entails a baseline equal in length to the radius of the Earth's orbit: one astronomical unit. A simple mathematical rule translates the measured stellar parallax **p** into stellar distance **d**, in astronomical units:

$$\mathbf{d} = 206{,}265 \ / \ \mathbf{p}$$

This relationship holds true only if the parallax angle **p** is expressed in units of *arcseconds*. An arcsecond is a mere $^1/_{3,600}$ of a degree, and it turns out to be an appropriately sized unit in which to express tiny stellar parallaxes. A somewhat larger unit, the *arcminute*, is $^1/_{60}$ of a degree, or sixty times larger than an arcsecond. (In geometry, these units are called *seconds* and *minutes;* astronomers have adopted the *arc-* prefix to avoid confusion with units of time.) For example, a star whose parallax measures one arcsecond lies 206,265 astronomical units away from the Earth. Precisely how many stars outside the solar system fall within this extraordinarily large distance? None! As the parallax hunters in this book were to discover, to their chagrin, stars in our galaxy are spaced unimaginably far apart. John Updike had it right when he wrote in *The Poorhouse Fair,* "The chief characteristic of the universe is, I would say, emptiness. There is infinitely more nothing in the universe than anything else."

So far we've considered a star's behavior only when viewed from two extremes of the Earth's orbit. It's as though we had taken a snapshot of the star at one moment in time, then again six months later, and concluded that the star flits from one spot in the heavens to another. But what happens to the star between these "snapshots"? Where does it appear in the sky, say, *three* months after we first observe it? And were we to monitor the star's position continually during the course of an entire year, what sort of figure would the star appear to trace in the heavens?

These are questions parallax hunters have asked themselves over the centuries; no astronomer on the trail of stellar parallax observes a star, then closes up shop for six months while awaiting a second look. On the contrary, astronomers record as many observations of a target star as they can, as frequently as possible. They expend every effort to tease out even the slightest hint of parallax, for this telltale angular shift is incredibly tiny, easily lost amid the turbulent blob of light that is the star in the eyepiece of a telescope.

As scientists, astronomers are skeptics by training, and often by nature. They are the proverbial Missourians who demand to be shown in the most unequivocal terms the validity of an observation before they are inclined to believe it. In the case of stellar parallax, two observations that purport to show a star first in one place, then in another, would be dismissed outright. There are plenty of alternative explanations to account for the star's "jumpy" behavior, most having to do with the incompetence or inexperience of the observer. To be convinced of the reality of an effect as subtle as stellar parallax, astronomers demand to see a series of well-timed and well-executed observations. They want to see data that has been corrected for the vagaries of the Earth's atmosphere, through which all starlight is filtered, and for the flaws and inherent peculiarities of the observer's telescope. They want to see patterns in the data that agree with what is theoretically expected. In short, they want to be convinced that parallax, and only parallax, can explain the body of evidence before them.

Specifically what might a parallax astronomer expect to see when studying a star continuously over the long term? The answer depends on the position of the star relative to the plane of the Earth's orbit, as shown in the illustration on the following pages. Let's consider three stars: Star A lies in the same plane as the Earth's orbit (which again appears foreshortened), star B lies perpendicular to the plane of the Earth's orbit, and star C lies in a position intermediate between A and B.

As the Earth circles the Sun, star A will appear to jog side-to-side along a straight line; whichever way the Earth swings in its orbit, the star appears to swing the opposite way, a reverse reflection of our planet's movement. To convince yourself that this is indeed the case, hold a finger upright again in front of your face and alternately blink your eyes. When you open your right eye only, your finger appears to shift to the left; when you open your left eye only, your finger appears to shift toward the right. The same effect holds true for the star as we view it from the orbiting Earth. Unlike the discrete jumps of your finger, however, the star's side-to-side oscillation in the sky is continuous.

Star B is situated along a line that is perpendicular to the Earth's orbital plane. As the Earth moves, the star appears to execute a tiny

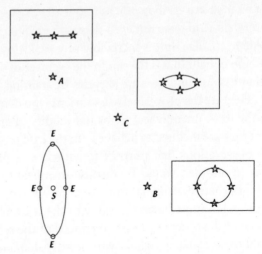

Annual parallactic movement of stars in different positions relative to the Earth's orbit.

circle in the heavens, again a reverse reflection of the Earth's orbital movement. Wherever the Earth happens to be in its orbit, the star will appear diametrically opposite in its own tiny parallactic circuit in the sky. To understand why this occurs, imagine an improbably long pencil whose eraser-end is affixed to the Earth and whose body is held loosely at the star. As the revolving Earth carries around the pencil's base, the point traces out a reverse circle in the heavens. The diameter of the star's parallactic circle depends on how far away the star is: The farther the star, the smaller its circle. If the star is sufficiently remote, then the circle is too tiny to be detected.

Star C lies at some odd angle, positioned between stars A and B. Its parallax trace is intermediate between star A's straight line and star B's circle: a flattened circle, or ellipse. The closer the star's position to the Earth's orbital plane, the flatter the ellipse. The straight-line oscillation of star A and the circular motion of star B can be considered extreme cases of the general elliptical path.

Stellar parallax is a by-product of the Earth's orbital motion. An imaginary observer on the Sun would see no parallax of the stars because the Sun is fixed at the hub of the solar system. Likewise, an observer on a remote planet, such as Pluto, would see much larger stellar parallaxes than we do from the Earth. Pluto is about forty times the Earth's distance from the Sun, so all parallaxes would appear forty times larger. That's the good news. The bad news is that Pluto has virtually no oxygen and a surface temperature 220 degrees centigrade *below zero,* so before we contemplate erecting an observatory on Pluto . . .

There's yet another factor that must be considered in predicting what the parallax hunter might expect to see when tracking a star for a long time. Up until now we've assumed that the star remains fixed in space while the Earth executes its annual revolution about the Sun. But the "fixed" stars, as we'll come to learn, do in fact move through space. As a result, while a particular star is coursing around its parallactic ellipse, it is simultaneously barreling through the void. The overall path perceived by an Earth-based observer is therefore a combination of the star's straight-ahead space motion and its elliptical parallactic motion.

To envision what such a hybrid path might look like, let's join the neighborhood kids for a nighttime game of flashlight tag. You're "it." The aim is to run after opponents and "freeze" them in the beam of your flashlight; "frozen" players must stand still until "thawed" by another player's beam. Suppose that a frozen player is marking time by slowly moving her flashlight in a small ellipse. Looking from a distance, all you see is the beam of the flashlight tracing out a compact loop in the darkness. Suppose further that the player is "thawed" and starts to run, while still moving the flashlight in the little ellipse. Now the beam's apparent path is a combination of the player's slow, looping arm movement and her straight-away running movement. In other words, the beam's former elliptical path is obliterated, stretched out before it has a chance to loop back on itself. The net result: a gleaming spot that weaves up and down as it progresses forward.

And that's precisely the path sought by the parallax hunter in watching a nearby, moving star over time. Patience is required, for it takes at least a year's careful observations to reveal beyond doubt the telltale sign: a straight-ahead trace with a slight, almost imperceptible, undulation. A speck of light advancing across the heavens, stepping ever so slightly off track, first in one direction, then in the other.

Imagine a tapestry, thick with interwoven threads, looping in, looping out. Now imagine the tapestry's threads fading into invisibility. Except for one. A lone thread, its undulating form preserved by unseen forces. A star slides along this thread like a bead along a necklace, following the thread's gentle arc as it meanders ever so slightly from side to side. An astronomer sights the star in a telescope, sees it oscillating—perhaps. It's hard to be sure. The star's path is reminiscent of the weave of a tapestry thread, but a weave of exquisite subtlety. A weave at the very limit of perception.

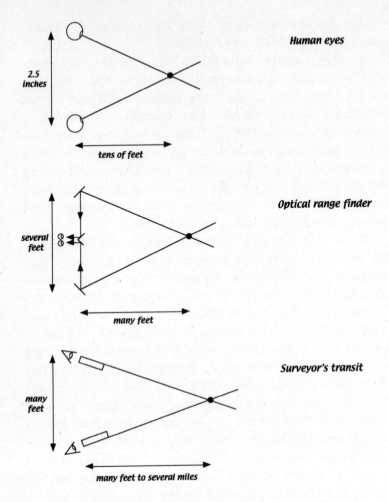

Human eyes

2.5 inches

tens of feet

Optical range finder

several feet

many feet

Surveyor's transit

many feet

many feet to several miles

Broadening the baseline: parallax measurement from near to far.

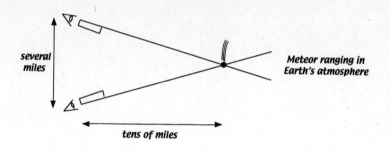

several miles

tens of miles

Meteor ranging in Earth's atmosphere

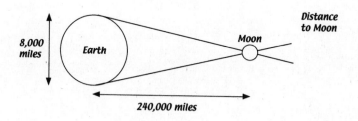

8,000 miles

Earth

Moon

Distance to Moon

240,000 miles

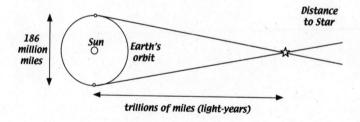

186 million miles

Sun

Earth's orbit

Distance to Star

trillions of miles (light-years)

⋆Part 2

Portrait of Tycho Brahe, originally from his 1596 book, *Epistolae*.
Source: *Owen Gingerich.*

5

The heavens Erupt

heaven's utmost deep
Gives up her stars, and like a flock of sheep
They pass before his eye, are number'd, and roll on.
—*Percy Bysshe Shelley, "Prometheus Unbound,"*
about the astronomer's work

I was so astonished at this sight that I was not ashamed to
doubt the trustworthiness of my own eyes. . . . A miracle indeed,
either the greatest of all that have occurred in the whole range of
nature since the beginning of the world, or one certainly that is to
be classed with those attested by the holy Oracles, the staying of
the Sun in its course in answer to the prayers of Joshua, and the
darkening of the Sun's face at the time of the Crucifixion.
—*Tycho Brahe, on the sudden appearance of a new star*
on November 11, 1572, from De Nova Stella

As a veteran astronomer, I've come to grips with the vastness of
space. The light-year, a distance measure that comprises a virtually
uncountable succession of miles, is a cosmic footstep to me. I'm
able to sweep my mind's eye through the Orion nebula, along our
Galaxy's spiral arms, then into—and out of—a black hole without
breaking a sweat. I can hold a good chunk of the universe in my
head all at once, and what I picture looks as cozy and ordered as
the Earth globe that sits on my desk. My vantage point on the universe is some imaginary cosmic overlook, impossibly high and far
away, yet accessible at a moment's reflection. I've even dabbled in
visualizing the infinite, although by definition it can't be done.

Large is fun. Maybe that's why astronomers, in my view, have
always seemed a contented lot. When astronomers think "large,"

they don't think elephant-large or skyscraper-large or even moun-tain-large; they think star-large, galaxy-large, universe-large. Popular astronomy books are peppered with words like immense, far-flung, vast, gargantuan, enormous—and with prefixes like super-, mega-, and hyper-. Bigness is part of the mindset of the astronomer. So it was natural for me to believe I was on intimate terms with the meaning of *large,* that it was so deeply rooted in my way of thinking that nothing—certainly nothing on the Earth—could impress me with its size. Then I saw the Grand Canyon.

It's almost three years now since I first stepped up to the Grand Canyon's South Rim. Driving into the park on that sun-drenched morning, I got no cues as to what lay ahead; Nature has hidden the chasm discreetly behind low pines and scrub. Even walking from the parking lot toward the rim, I saw nothing through the veil of vegetation. Suddenly I emerged into the open and stopped. I didn't know where to look because the canyon was everywhere. I had read about the Grand Canyon, seen pictures of it, plumbed its depths on a topographic map. The canyon unrolled into the distance with such velocity that I almost sensed a tug sweeping me forward.

When I look at the sky, as I do practically every night, I'm looking *up.* Perhaps it's because my head is thrown back that I know what I'm seeing is elsewhere, away from the Earth, away from me. But the Grand Canyon stretched away from my feet in an aggressively horizontal thrust. In one continuous take, I could run my eyes from the tips of my sneakers all the way to the canyon's far rim. I felt my eyes rapidly shift focus as they flitted among the near rim, the far rim, and all the myriad ridges in between. I tried to comprehend in a logical way what I was seeing, but logic is the poor stepchild of awe. So I just stood there and looked until my kids pulled me away.

According to the tourist brochure, the canyon's North Rim is about a ten-mile beeline from where I was standing (although it's a five-hour, 215-mile roundabout trek to get there by car). The air was clear that day and, squinting against the sunlight, I could just make out the Grand Canyon Lodge perched on the North Rim. Strolling a short way along the South Rim, I noticed that the lodge always appeared in the same position against the background landscape, even though every step gave me a slightly different vantage point on it. In other words, the lodge displayed no detectable parallax over the baseline defined by my stroll.

Astronomers up to the sixteenth century were in a similar boat; despite their best efforts, they had been unable to detect the parallax of a star. From this, most of them concluded that the Earth was immobile in space. By the same reasoning, the absence of parallax for the Grand Canyon Lodge would be taken to imply that I had remained rooted in one spot, like the immobile Earth of Ptolemy. Yet I *had* moved. The parallax that would have proved my change in position was simply too small to be detected by eye. This was the challenge faced by Copernicus's followers, who held that the Earth orbits the Sun and thus alters their vantage point on the stars. The ultimate proof of the Copernican system, they believed, hinged on the detection of stellar parallax. Generations of Copernican astronomers went to their graves believing that the measurement of stellar parallax lay right around the corner. With the benefit of hindsight, we know just how wrong they were and just how difficult the task really was.

Let's scale down the solar system and its neighboring stars so they fit into the Grand Canyon. My position on the South Rim represents the Earth; the lodge on the North Rim ten miles away represents the *nearest* star beyond the Sun (in reality, about 27 trillion miles distant). On this scale, the 186-million-mile width of the Earth's orbit reduces to a measly four inches. Detecting a star's parallax from the extremes of the Earth's orbit is like detecting the parallax of a ten-mile-distant object over a four-inch baseline. And that's for the nearest star. Other stars would be harder.

Stellar parallax—and hence stellar distance—could not be measured until the groundwork had been laid. At the time Copernicus published his heliocentric theory in 1543, practical astronomy was in a sorry state. Most of its few practitioners carried out random observations with marginal equipment that they used carelessly. Someone had to transform the practice of astronomical observation into the *science* of astronomical observation. In the latter half of the sixteenth century, there was only one person in all of Europe who seemed up to the task, an irascible, red-bearded Danish nobleman named Tycho Brahe. Tycho was prepared to devote his entire life to improving astronomy. That is, if he could avoid getting himself killed first.

Jörgen Brahe and his wife longed for a child of their own—a boy, preferably. They would raise the boy in the privileged lifestyle of the landed Danish gentry. They would see to their son's every need and want, send him to the finest university, make him a worthy successor to the Brahe name and fortune. But as the sixteenth century neared its midpoint, God had not favored Jörgen and his wife with a baby. Jörgen was growing desperate.

Jörgen's brother, Otto Brahe, counseled patience, and in a fit of remarkable generosity promised the childless couple an eventual child—one of his own. Thus, the birth of Otto's first son, Tycho, on December 14, 1546, at 10 o'clock in the morning, was celebrated in two Brahe households. The sight of the baby must have awakened Otto to the folly of his contract with his brother, for he summarily withdrew his offer. Just over a year later, a second son was born. With a "spare" son, as it were, in Otto's household, Jörgen decided it was time to enforce Otto's promise. He stole into Otto's home and absconded with Tycho. In the end, the brothers Brahe came to an accord in their struggle over the baby. Tycho would be raised by his uncle and aunt.

In young Tycho, Jörgen saw a future statesman. He had Tycho educated by private tutors, and in 1559 enrolled his thirteen-year-old "son" at the University of Copenhagen to study philosophy and rhetoric. In his extracurricular hours, Tycho practiced his favorite hobby: astrology and prognostication. So Tycho was primed for the event that was to alter his life. On August 21, 1560, a solar eclipse occurred. The eclipse appeared total to residents at the latitude of Portugal, but far to the north in Denmark, only a portion of the Sun's face was obscured by the Moon. A partial eclipse isn't very impressive; it's easy to miss it entirely unless you know ahead of time that it's due to occur. Despite their "creakiness," both the Ptolemaic and Copernican models had foreseen the August eclipse of 1560. Tycho stood outside at the appointed hour and, as if on cue, the Moon's dark form began to intrude upon the Sun. It was not so much the eclipse itself that impressed Tycho, but its prediction. How could astronomers foretell the occurrence of a heavenly event? How had they inserted themselves into the mind of God? Shortly afterward, Tycho bought his first astronomy book. By year's end, he was plowing through a Latin translation of Ptolemy's works. Uncle Jörgen was not pleased. Astronomy was not a proper pursuit for a Danish nobleman. Tycho might as well become a swineherd.

In 1562, Jörgen transferred his headstrong nephew to the University of Leipzig. Perhaps there he would focus on essential studies, specifically the law, not a triviality like astronomy. But just in case, Jörgen hired a *Hovmester*—a companion—to keep Tycho in check. Anders Sörensen Vedel, a history student at Copenhagen, had come highly recommended. Although only four years older than Tycho, Vedel took his job seriously. When Tycho expressed an interest in mathematics, Vedel dutifully reminded him how

Copenhagen's mathematician, Ejler Hansen, had gone insane from studying the subject too much. Tycho enrolled anyway.

Vedel doled out Tycho's allowance—which Tycho surreptitiously spent on astronomy books and instruments—and kept a stern watch for any deviation from Uncle Jörgen's rules. But every night, after Vedel had gone to bed, Tycho pulled out his fist-sized celestial sphere and his books from their hiding place and taught himself the ways of the cosmos. This cat-and-mouse game stirred up a lot of tension between the two young men. Eventually, however, Vedel tired of his role as Tycho's keeper. The boy might be self-centered and quick to temper, but Vedel had seen his share of spoiled noblemen's sons who hadn't a fraction of this one's enthusiasm for learning. The sphere, the books, the instruments came out of hiding. Tycho and Vedel became lifelong friends.

By the time he was sixteen years old, Tycho had taught himself how to predict planetary positions and eclipses according to both the Ptolemaic and Copernican systems. Tycho's first recorded series of observations began on the night of August 17, 1563, with the impending conjunction—close passage—of Jupiter and Saturn. Tycho was a firm believer in astrology; conjunctions were considered especially useful in casting horoscopes and predicting the future. (His predictive record was spotty: He once publicly forecast Suleiman's imminent death, only to be informed that the Ottoman sultan had been killed almost two months earlier in an attack on Hungary.) To keep track of the separation between Jupiter and Saturn in the sky, Tycho used the only angle-measuring device he had: a V-shaped draftsman's compass. Holding the apex of the compass near his eye, he spread the two compass legs until one pointed toward Jupiter, the other toward Saturn. Then he laid the compass flat on a piece of paper, on which he had already drawn a circle divided into degrees, and read off the angle between the two legs. Night after night for several weeks, Tycho measured the narrowing gap between Jupiter and Saturn. On the evening of August 24, the two planets were so close to one another in the sky that they appeared almost as a single bright speck.

As beautiful as this celestial sight must have been, Tycho was appalled. The predicted conjunction date according to the Ptolemaic model was a month off, that of the Copernican model a day off. Tycho realized what other astronomers apparently had not: Improvements in prediction would come only after astronomers carried out continuous, decades-long, high-precision observations of stars and planets. Observed planetary positions were the fuel required by the mathematical engines of the cosmic models; if there was not enough fuel or if the fuel was of poor quality, the engine would never run right. Tycho's goal was nothing less than the reinvention of observational astronomy.

In June 1565, Uncle Jörgen died unexpectedly. (He'd contracted pneumonia after rescuing the Danish king, Frederick II, from drowning when the king fell from his horse off the Amager Bridge near the royal castle in Copenhagen. Both Frederick and Jörgen might have been drunk at the time.) Jörgen had not fully completed arrangements to make Tycho his legal heir, so upon his death his entire estate went to his wife. Nineteen-year-old Tycho returned to the home of his biological parents for a time, but he grew restless. He felt driven to become an astronomer, despite the inevitable disapproval of relatives and friends. In the spring of 1566, Tycho embarked on a search of astronomers who might have something to teach him. He found mentors at the universities in Wittenberg, Rostock, Basel, and Augsburg.

Tycho's stopover in Rostock was eventful for a nonacademic reason. On December 10, 1566, he attended a party at the home of theology professor Lucas Bachmeister. There he got into a heated argument with another Danish nobleman and distant relative, Manderup Parsbjerg. One report claims that the dispute centered on which of the two men was the more able mathematician; another that Parsbjerg mocked Tycho's after-the-fact prediction of Suleiman's death. Whatever the reason for their animosity, the argument resumed at a Christmas celebration several weeks later, then culminated in a duel on the evening of December 29. (The Brahe clan, like most Danish nobles of the time, were a hot-headed lot: One of Tycho's third cousins killed a man in a duel in 1568, an uncle killed a second cousin in 1581, another third cousin killed a second cousin in 1584, and Uncle Jörgen's brother was killed in 1592.) Tycho's duel ended abruptly when Parsbjerg lopped off a sizable portion of Tycho's nose. During his recovery, Tycho must have reflected on the impact the disfigurement would have on his life. Metallurgy being one of his many skills, Tycho fabricated his own prosthesis of gold, silver, and copper, tinted to match his own skin. (Doubts about the tale of Tycho's nose led Prague city officials to open Tycho's marble tomb in 1901. Two physicians from the medical school examined Tycho's skeleton, still clad in a red, silk robe. They found "a narrow, curved mark on the skull at the upper end of the nasal opening" that was consistent with a sword wound; the opening was "rimmed by a bright green stain of copper." The examiners concluded that Tycho

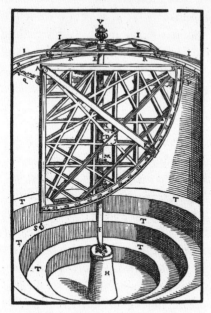

QVADRANS VOLVBILIS
AZIMVTHALIS.

One of Tycho's large quadrants. From
Tycho's *Astronomiae Instauratae
Mechanica*, 1602.
Source: *Widener Library, Harvard University.*

had indeed worn a metal prosthesis. What became of it is unknown.)

In 1569, twenty-three-year-old Tycho received a commission to construct a world-class instrument: a colossal quadrant for the garden of wealthy Augsburg alderman and amateur astronomer Paul Hainzel. A quadrant is a 90-degree section of a protractor and is used to measure elevations of celestial objects. Tycho made the quadrant nineteen feet tall, large enough that he could easily subdivide each degree measure into sixty equal parts, that is, into arcminutes: altogether 5,400 hairline marks along the quadrant's brass-plated curve. Twenty men were needed to raise the finished quadrant's massive oak frame in Hainzel's garden. The quadrant sat upright against an oak pillar, such that the entire assembly could be rotated to access any part of the sky. The Augsburg quadrant established Tycho's reputation as an astronomical instrument designer. (The quadrant was destroyed in a storm after only five years of service.)

After his father's death in 1571, Tycho returned again to Denmark. He moved to the secluded estate of his maternal uncle, Steen Bille, at Heridsvad Abbey, a former Benedictine monastery. Bille gave his nephew permission to set up a chemical laboratory in an outbuilding on the grounds. Over the next year, Tycho observed the heavens infrequently, choosing instead to develop his skill in chemistry. That changed abruptly on the night of November 11, 1572. Tycho had just closed up his laboratory for the evening and was crossing the lawn to the main house. He gazed at the sky while he walked (the same habit that led his ancient predecessor Thales to stumble into the well over 2,000 years before). The identical stellar patterns that had glittered above Aristarchus, Hipparchus, and Ptolemy now shone down on Tycho. The night sky was as familiar to Tycho as your own neighborhood is to you. But on this particular night, Tycho noticed with amazement, the sky had changed. In the constellation of Cassiopeia was a new star, a magnificent star almost as bright as the planet Venus. Tycho tells the story in his book *Progymnasmata*, published posthumously in 1602:

> *Behold, directly overhead, a certain strange star was suddenly seen, flashing its light with a radiant gleam and it struck my eyes. Amazed, and as if astonished and stupified, I stood still, gazing for a certain length of time with my eyes fixed intently upon it and noticing that the same star placed close to the stars which antiquity attributed to Cassiopeia. When I had satisfied myself that no star of that kind had ever shone forth before, I was led into such perplexity by the unbelievability of the thing that I began to doubt the faith of my own eyes, and so, turning to the servants who were accompanying me, I asked them whether they too could see a certain extremely bright star when I pointed out the place directly overhead. They immediately replied with one voice that they saw it completely and that it was extremely bright. But despite their affirmation, still being doubtful on account of the novelty of the thing, I enquired of some country people who by chance were travelling past in carriages whether they could see a certain star in the height. Indeed these people shouted that they saw the huge star, which had never been noticed so high up. And at length, having confirmed that my vision was not deceiving me, but in fact that an unusual star existed there, beyond all type, and marvelling that the sky had brought forth a certain new phenomenon to be compared with the other stars, immediately I got ready my instrument.*

Although Tycho described the object as a *nova stella*, a generic term for "new star," we believe he had witnessed a *supernova*, the titanic explosion of a dying star. By contrast, the ancient observer Hipparchus had probably viewed what modern astronomers term a nova, a temporary, nondestructive eruption on the surface of a very hot star. Hipparchus's star, although it quickly dimmed below visibility, likely survives to flare another day. Tycho's star, on the other hand, blew up; its gaseous debris, by now scattered over 250 trillion miles of space, will mix with other pockets of interstellar gas and together condense into future generations of stars.

Tycho was aware of Hipparchus's long-ago account of a new star in the heavens, but most sixteenth-century astronomers believed Hipparchus's "star" had in fact been something else, perhaps a comet without a tail. So by Tycho's day, it was still canon that the sphere of the stars was immutable; whichever stars existed there at the Creation would never be extinguished, nor would any new stars join them. Yet plainly visible above Tycho's head on that crisp November night blazed proof to the contrary. The wisdom of the ages, it appeared, was wrong. As to Tycho's emotion upon seeing that beacon in the sky, simply imagine how you would feel if you looked out your window and saw a unicorn grazing on your front lawn.

Tycho's mind must have been filled with questions. Will the new star shine the following night? Will its brightness change? Will it move? He drew up a plan of action. The first order of business was to keep track of the star's position from night to night. For this, he used a five-foot walnut sextant to record the angles between the new star and nine surrounding stars in Cassiopeia. The sextant (not to be confused with a mariner's sextant, which measures altitude, not the angle between celestial objects) was similar to the giant quadrant he'd designed for Hainzel in Augsburg, but spanned only 60 degrees. Attached to the sextant's arc was a bronze strip divided into degrees and arcminutes, or sixtieths of a degree. At the far ends of the sextant's two arms were peepholes through which pairs of stars could be sighted.

For the new star, Tycho fashioned a new astronomy. Gone were the haphazard observations, impressionistic descriptions, and sloppy methodologies of the past. Over the succeeding months, Tycho kept a scrupulous record of the star's position, brightness, and color. He repeated observations multiple times each night, cross-checked the results from various instruments, and compensated for instrumental flaws. His goal was to reduce errors and uncertainties in his measurements as best he could and to leave a complete, quantitative chronicle of this remarkable event for future generations of astronomers. For if the celestial sphere changed in 1572, it might someday again. (In fact, just thirty-two years later, in 1604, another supernova erupted; this explosion was observed by Tycho's eventual successor, Johannes Kepler. The next naked-eye supernova after that—and the first since the development of the telescope—did not appear until 1987.)

The supernova of 1572 remained visible for eighteen months. Initially it could be seen in broad daylight and sometimes even through clouds at night. It still rivaled the brightest stars in the sky half a year later. Over the months, its color changed gradually from white to yellow, then to red and finally to a leaden gray. Then it faded from view. To Tycho, the burning question was this: Where was the new star situated in space? Was it indeed a full-fledged star that had burst onto the celestial sphere? Or was it merely a luminous discharge—a tailless comet, perhaps—from combustible gases in or above the Earth's atmosphere? In short, was the new object far away or relatively nearby?

There was only one way to tell the object's distance for sure: detect its parallax. Tycho knew that during the course of the night, the sweep of the Earth's rotation gave him an ever-changing vantage point on the heavens. If the new object was closer than the Moon, then over the course of hours it would shift in the sky relative to the faraway stars. On the other hand, if the object was among the stars themselves, it would show no such shift. After months of careful measurement, Tycho had found no hint of parallax whatsoever. He concluded that "this star is not some kind of comet or fiery meteor, whether these be generated beneath or above the Moon, but that it is a star shining in the firmament itself—one that has never previously been seen before our time, in any age since the beginning of the world."

At a social gathering in Copenhagen in early 1573, Tycho described his ongoing observations to a friend, Professor Johannes Pratensis. Although Pratensis had read Pliny's report of Hipparchus's star, he scoffed at the idea that such an occurrence could happen again. So did French envoy Charles Dancey, who suspected Tycho of playing a joke. Tycho invited the guests outside and, to their amazement, pointed out the new star. Pratensis urged Tycho to publish his results. Tycho refused. It was one thing for a Danish nobleman to practice in private an "ignoble" subject like astronomy, quite another to publicize his involvement. However, a few months later, reports about the new star began to filter in from Germany. Most claimed the object to be a nearby comet. The German painter Georg Busch weighed in with his alternative explanation that the star was a comet "formed by the ascending from earth of human sins and wickedness, formed into a kind of gas, and ignited by the anger of God. This poisonous stuff falls down again on people's heads, and causes all kinds of mischief such as pestilence, Frenchmen, sudden death, bad weather, &c."

Tycho couldn't let such nonsense go unchallenged. He released a slim, hundred-page book known today as *De Nova Stella*, "On the New Star," a merciful contraction of the full eighty-five-word title. Here Tycho gives a sober account of his observations of the new star, in contrast to the absurdities put forth by the "common herd of scribblers." Tycho explains that the new star's lack of parallax placed it unquestionably in the celestial sphere, contrary to centuries of collective wisdom that the heavens do not change. Tycho concludes *De Nova Stella* with an eight-page elegy declaring his intention to continue his work in astronomy despite his nobleman's status.

Basking in the acclaim brought by his book, Tycho resumed his travels through Europe. "An astronomer," he wrote, "more than the students of other branches of knowledge, has to be a citizen of the world, and consider every place to which circumstances or necessity might lead him as his native country." On this swing across the continent, Tycho sought a patron to support what he saw to be his life's calling: to remedy the observational deficiencies that hobbled the progress of astronomy. Among those he met was astronomer and statesman Landgrave (Baron) Wilhelm IV of Hesse in southwest Germany. Wilhelm was so impressed with Tycho that he subsequently contacted the Danish king, Frederick II, urging him to provide means for the talented Tycho to pursue his astronomical research. Otherwise, Tycho would certainly forsake Denmark for better prospects elsewhere: "Your Majesty must on no account permit Tycho to leave, for Denmark would lose its greatest ornament."

With his nation's pride at stake, Frederick dispatched a courier to summon Tycho. The rider was instructed to travel day and night until he reached the Brahe family estate at Knudstrup, across the sound from Copenhagen. As Tycho later recounted in a letter, he was lying awake in bed, ruminating about his future, when the courier (a relative of Tycho's) burst into the room two hours before dawn and pressed the king's summons into his hand. Tycho dressed immediately, rode toward the royal hunting lodge outside Copenhagen, and arrived that evening. There the king made an offer Tycho couldn't refuse—money to establish and equip a Danish observatory, an annual stipend to fund its operation, and, finally, an entire island on which to build it.

On February 22, 1576, Tycho stepped off a boat onto the island of Hven in the sound between Denmark and Sweden. Hven's white cliffs rise abruptly from the bay, then level off to a wide, grassy plateau. Altogether 2,000 acres, deeded to Tycho for life. There was a single village at Hven's northern end called Tuna (derived from the Scottish word for "town") and communal farmland and pastures, several stands of trees, a church, and the obligatory windmill. From anywhere on the island, Tycho could see the wooded coastlines of Denmark on one side and present-day Sweden, then part of Denmark, on the other. Nine miles north across the water lay Elsinore, the eventual setting for a tragedy about a Danish prince by a playwright named William Shakespeare, at the time just twelve years old. (Curiously, the names "Rosenkrans" and "Guldensteren" appear among the ancestral coats of arms surrounding the portrait of Tycho that opens this chapter.)

Just before dawn on August 8, 1576, with Jupiter beaming in the east and the full Moon in the west, Tycho stood on the spot he had selected for his new home and observatory. With the help of his common-law wife, Christine, and several friends, he laid the

Painting of Uraniborg by Heinrich Hansen, 1882.
Source: *Det Nationalhistoriske Museum på Frederiksborg, Hillerød, Denmark.*

foundation stone. The observatory would be called Uraniborg, after Urania, the Greek muse of astronomy. Construction lasted five years.

Tycho spared no expense in the execution of his plans. Inspired by Palladio's Villa Rotonda near Vicenza, Italy, Uraniborg was a place Euclid would have loved, a tantalizing confection of intersecting circles, arcs, and squares, symmetrical to a fault. Every brick, every board, every bush had its place. Tycho created more than quarters in which to live and work. He transformed the island of Hven into his personal well-ordered universe: a "Tycho-centric" universe whose motive force was himself. In the middle of the Uraniborg complex stood the house, which combined residence and observatory. Surrounding the house, in turn, were a circular courtyard, a square ornamental garden, and a border of trees. The entire arrangement was enclosed by four stone-covered earthen walls, each 248 feet long, 18 feet high, and 16 feet thick at the base. The enclosure's corners were precisely aligned to the points of the compass. There were access gates at the east and west corners, above which were kept English mastiffs, canine "doorbells" whose barks alerted Tycho of every visitor's arrival. Among the many visitors was Tycho's "watchdog" from his Leipzig days, Anders Vedel, who had since become Denmark's Royal Historian. Buildings were located at the north and south corners, one a printing office, the other housing for the domestic staff. A small prison was conveniently situated in the basement of the latter. (Tycho was not only landlord of the tiny principality of Hven but also judge and jury.) Fish ponds dotted the landscape beyond the walls.

Uraniborg itself was constructed in a Gothic Renaissance style of red bricks with ornamental sandstone and limestone. Spires jutted upward from the copper roof, and there was a profusion of decorated gables and cornices. The main structure itself was three stories tall and about fifty feet on a side. To the north and south were attached four round towers, two on each end, the largest of which was eighteen feet across. Within these towers were housed the astronomical instruments. The towers' triangular roof boards could be removed to gain access to the night sky. Outside galleries ran the perimeter of the house to accommodate portable instruments. The entire structure was surmounted by a domed octagonal clock tower, from which rose a weathervane in the shape of Pegasus, the mythological winged horse.

Inside were living quarters for Tycho and his family; guest rooms, a study, and a kitchen; a library with a five-foot-diameter, brass-sheathed celestial globe on which Tycho marked every star he observed (the globe was destroyed in a fire at the University of Copenhagen in 1728); aviaries; a chemical laboratory; and, in the middle of it all, a rotating fountain. (Yes, the house had piped running water.) Under the eaves were tucked eight unheated garrets for students: an astronomical boarding school. Tycho could summon any one of his charges by pulling on a set of strings embedded in the walls; each string rang a bell in a particular garret. He used to impress visitors by tugging secretly on a given string while whispering a student's name, only to have that student appear moments later as if by magic. Like a monarch, Tycho had his own jester, a dwarf named Jeppe, who sat at Tycho's feet during meals and was tossed an occasional morsel. According to the diary of one of Tycho's students, Jeppe "chattered incessantly, and . . . was supposed to be gifted with second-sight, and his utterances were therefore listened to with some attention." Whenever Tycho was away on business, Jeppe eagerly awaited his master's return, and when he spied him getting off the boat shouted, "*Junker paa Landet!*" ("The squire is on the land!")

The island of Hven was run like a feudal estate. No one was permitted to cut wood or gather nuts without Tycho's permission. A tenant who did not keep his fence in good repair had to pay Tycho a fine and provide a barrel of beer to the townspeople. Court sessions were held every second Wednesday. Tycho "was an oppressive landlord to the farmers on his island," according to astronomer Joseph Ashbrook. "To many of his contemporaries he must have seemed more a hot-tempered country squire than one of the greatest scientists of the time."

In 1584, Tycho built an adjunct observatory on a small rise outside the walls. Despite its soaring name—Stjerneborg, or "Star Castle"—this facility was subterranean; only its protective domes showed above ground. The instruments themselves sat in bunkers, sheltered from the winds that sometimes raked the island. On the walls of Stjerneborg's main chamber were portraits of famous astronomers, including Tycho himself, and an imagined rendering of Tycho's yet-to-be-born descendant, Tychonides.

Tycho hammered another nail into the coffin of Aristotelian beliefs during his second year on the island, while Uraniborg was still under construction. On November 13, 1577, almost five years to the day after he first saw the supernova of 1572, Tycho was fishing at one of his many ponds. He noticed in the darkening sky a bright, hazy patch near the head of Sagittarius the archer. As the sky dimmed further, he could make out a luminous tail that extended more than 20 degrees from the nucleus of light: a comet. (At about the same time, a six-year-old Johannes Kepler was escorted up a hill in Germany to view the heavenly wonder.)

Aristotle had held that comets were combustible gases in or just above the Earth's atmosphere. Tycho knew that if this were the case, then during the course of the night, a comet would display a parallactic shift against the much more distant stars. Tycho and his assistants observed the comet until it faded from sight at the end of January. No parallax was detected. As a check, Tycho collected comet measurements from other astronomers and proved that on any given night the comet appeared in the same position for every observer. Again, no parallax. The comet of 1577, he concluded, existed beyond the Moon, somewhere among the planets. His 1588 treatise on the subject obliterated the long-held Aristotelian theory of comets. Systematic observation and analysis had triumphed over speculation.

In the astronomical world, supernovae and comets are rare occurrences. Tycho's bread-and-butter work—the work he built Uraniborg for—was the measurement of star and planet positions. Uraniborg's primary instrument was the six-foot brass mural quadrant, which was permanently mounted on an east-facing wall in one of the rooms of the main house. Starlight entered through a small hole in the south wall, and the Earth's rotation brought celestial objects one by one into view, at which time their altitudes would be called out and recorded. On the wall within the quadrant's arc was a life-size

QVADRANS MVRALIS
SIVE TICHONICVS.

Tycho's six-foot mural quadrant. From Tycho's *Astronomiae Instauratae Mechanica* (1602), reproduced in Schweiger-Lerchenfeld (1898). **Source:** *Widener Library, Harvard University.*

painting of Tycho himself seated at a table and pointing to the sky, his dog at his feet; behind him is a rendering of Uraniborg's various observatories, abuzz with activity. At these locations, Tycho used quadrants, sextants, armillary spheres (a series of intersecting, graduated rings along whose diameters celestial objects could be sighted), and an eight-foot-long device called a *triquetrum*, made by Copernicus himself.

Science writer Willy Ley paints a vivid picture of a typical working day at Uraniborg:

> *When [Tycho] measured the altitude of a star with his great quadrant he had one or two assistants, or pupils, help him. A third would sit nearby at a writing desk and note down the figures which Tycho shouted at him. In addition to assistants and pupils, Uraniborg was overrun with artisans and craftsmen, most of them German or Dutch, with Danish country folk tending the garden and Danish serving girls swarming in the kitchen. . . . Hven was supposed to be a quiet place to study the sky, but it had turned into a principality, which differed from others mainly in the fact that its lord—the men around Tycho called him Junker Tyge—chased star positions instead of stags or women.*

Tycho was a one-man national observatory. As a result of his research, Denmark became a worldwide power in observational astronomy. Tycho carried on a running correspondence with astronomers throughout Europe, exchanging information, providing collaboration, and promoting himself (rightly so). The bulk of the Uraniborg work was tedious by most standards. Yet the steady accumulation of accurate stellar and planetary position data was absolutely necessary for the advancement of astronomy. Cross-checking Tycho's star catalogue against those of more modern vintage, we know that he determined the places of stars to within one arcminute—$^1/_{60}$ of a degree—of their actual positions. An astonishing achievement, considering that he worked without the benefit of a telescope. Thus, Tycho increased the accuracy of star and planet positions up to tenfold from that of previous catalogues. His work was accurate enough to provide future astronomers with the raw data they needed to deduce the mechanics of planetary motion.

"It is a difficult matter," Tycho writes in *De Nova Stella,* "and one that requires a subtle mind, to try to determine the distances of the stars from us, because they are so incredibly far removed from the Earth." Tycho tried repeatedly to measure the parallaxes of stars. He never succeeded. As accurate as his instruments were for their time, they were still far too coarse for the job; stellar parallax simply cannot be measured by eye. In the latter half of the sixteenth century, the Ptolemaic world model still held sway, and Tycho—at least initially—was one of its adherents. The Copernican model, although generally better at predicting eclipses and planetary positions (but not a whole lot better), was the upstart competitor. In the absence of stellar parallax, Tycho faced the same choice as the rest of his contemporaries: to assert that the Earth is central and immobile and the stars are relatively near or that the Earth swoops around the Sun and the stars are so remote that their parallaxes fall below the threshold of measurement.

Tycho knew the capability of his instruments. He calculated that the stars must lie at least 700 times farther than Saturn, the outermost known planet at the time; otherwise he surely would have detected their parallaxes. But like so many others before him, Tycho fell victim to his preconceived notions about how the universe should be. Such an immense stellar distance—700 times that of Saturn—created an unimaginably large and, in Tycho's opinion, wasteful gap between the planets and the stars. (By modern measurement, the nearest star lies about 30,000 times farther than Saturn.) The lack of stellar parallax assured Tycho that Copernicus must have been wrong; the Earth must be fixed at the center of the universe. At the same time, Tycho was acutely aware of the deficiencies of the Ptolemaic system. Therefore, Tycho introduced his own extraordinary, lopsided world model in which the five heavenly planets orbit the Sun, while the Sun simultaneously orbits a central Earth: a "geocentric-heliocentric" hybrid—the *Tychonic* system.

The Tychonic system is mathematically equivalent to the Copernican system; thus, its predictions will be no better or worse. But one bold innovation was forced upon Tycho by the planet Mars. In his model, the orbit of Mars intersects the orbit of the Sun. Such a scheme seemed impossible if the heavenly bodies are affixed to solid crystalline spheres, as was still commonly believed. "I now no longer approve of the reality of those spheres," Tycho writes in his 1602 treatise *Progymnasmata,* "the existence of which I had previously admitted, relying on the authority of the ancients rather than driven by the truth of the matter itself. At present I am certain that there are no solid spheres in heaven, no matter if these are believed to make the stars revolve or to be carried about by them."

Thus, Tycho's model, although peculiar-looking from a modern perspective, made a bold conceptual advance: With the "shattering" of the age-old crystalline spheres, some other mechanism must hold the planets in their orbits. It would be another century before Isaac Newton discovered the nature of that mechanism: gravity.

Tycho's patron, King Frederick II, died in 1588. He was succeeded by his eldest son, Christian, who was only eleven at the time. The ruling council of four noblemen continued Tycho's support until 1596, when Christian assumed control of Denmark. In March 1597, the frugal Christian cut off all funding for Uraniborg. In a huff, Tycho packed up his instruments, library, printing press, and belongings and by month's end moved to his house in Copenhagen. Uraniborg fell into disrepair, and after Tycho's death was incrementally broken down for building materials. In 1623, a Hven resident paid a mason for 60,000 bricks "pulled down and renovated from the old castle at Oranienborg." By the mid-1600s, the great observatory building was gone; only its foundation hole remained, and that was soon covered over by grass. Today all that is left of Uraniborg are a few bricks on display in a museum.

Tycho stayed in Denmark only three more months, then moved his family, equipment, and an entourage of students first to Rostock, then Hamburg, Wittenberg, and Dresden. Finally, during the summer of 1599, Tycho resumed his astronomical research in Benatky Castle outside Prague, supported by the German emperor, Rudolph II. Early the next year, a new assistant arrived: the brilliant German mathematician Johannes Kepler, then twenty-eight years old. By this time, the fifty-three-year-old Tycho was largely spent, both physically and intellectually; nevertheless, he had sufficient energy to make Kepler's life miserable. In the words of novelist Arthur Koestler, "Tycho was an aristocrat, Kepler a plebian; Tycho a Croesus, Kepler a church-mouse; Tycho a Great Dane, Kepler a mangy mongrel. They were opposites in every respect but one: the irritable, choleric disposition which they shared."

Tycho and Kepler were the odd couple of 1600s astronomy. Despite their differences in station, appearance, and manner, they needed each other. Kepler knew that in Tycho's possession were the raw observations that he, as "architect," longed to assemble into a coherent picture of planetary motion. And Tycho knew that the gifted Kepler had the mathematical wherewithal to prove the validity of the Tychonic system of the heavens. But Kepler was a confirmed Copernican; Tycho's model had no appeal to him, and he had no intention of polishing this flawed edifice to the great man's ego.

The mass of planetary data drew both Tycho and Kepler with inexorable force. Were these two men a pair of explorers stranded in the desert, the data would have been the final, life-sustaining swallow of water in the canteen. Each man viewed the data as the means to his own salvation: to Tycho, that he might be immortalized as more than just a maker of observations, but a maker of worlds; to Kepler, that he might satiate his desire to learn the ways of the universe. When Kepler arrived in Prague, Tycho and an assistant were getting nowhere trying to make sense of the observations of the planet Mars. In Kepler's words, Tycho was "obstructed in his progress by the multitude of the phenomena and by the fact that the truth is deeply hidden in them." Kepler boasted that he himself would compute the precise size and form of Mars's orbit in a week's time; in the end, he would toil on the problem for five years.

The tumultuous partnership between Tycho and Kepler did not last long. On October 13, 1601, Tycho became feverish at dinner. He died eleven days later, with his family and Kepler in attendance. Just before he died, he is reported to have uttered, *"Ne frusta vixesse videar,"* "Let me not seem to have lived in vain." Tycho's legacy to Kepler—and through him to astronomy as a whole— was the wealth of accurate positions of the planet Mars. With these, Kepler ultimately deduced the crucial element that all his astronomical brethren had missed: The orbits of planets are ellipses, not circles. It's significant that Kepler so trusted Tycho's Mars data that he took the minuscule deviation of the resultant orbit from perfect circularity to be *real,* not the effect of errors in Tycho's observations. Without such accurate measurements, the slight ellipticity of Mars's orbit would have been impossible to tease out. Kepler went on to derive the fundamental laws of planetary motion, dispensing with the epicycles and other contrivances that marred the Copernican cosmos.

On May 4, 1600, a little over a year before he died, Tycho had posted a letter introducing himself and his new world system to a younger Italian colleague named Galileo Galilei. Galileo did not respond. Had the irascible Dane lived longer, he might have succeeded in making the acquaintance of the equally irascible Galileo, who was to use a new device—the telescope—to explore the heavens. Galileo would have approved of Tycho's scrupulous approach to astronomical research. And he would have made some useful suggestions on how to measure stellar parallax.

The earliest surviving depiction of spectacles: a 1352 fresco panel by
Tommaso Barisino of Modena.
Source: *Alinari/Art Resource, New York.*

6

The Turbulent Lens

A fool sees not the same tree that a wise man sees.
—William Blake, "Proverbs of Hell,"
The Marriage of Heaven and Hell, *1790*

Who, when he first saw the sand or ashes . . . melted in a
metallic form . . . would have imagined that, in the shapeless
lump, lay concealed so many conveniences of life?
—Samuel Johnson, 1750

In the movie *Zelig*, Woody Allen's character is seen clinking glasses with F. Scott Fitzgerald, greeting photographers at the arm of Josephine Baker, and being summarily shooed from a Vatican balcony by the Pope himself. Allen appears to parade through these scenes, not with actors, but with the historical figures themselves. We viewers wink at this cinematic sleight-of-eye. We know that the images are an illusion, that the events never took place. We disregard the evidence of our own eyes because our brain reminds us that Woody Allen was but a child when *Zelig*'s "newsreel footage" was supposedly shot. (And in any case, Woody Allen with the Pope? Not likely.) We are wise enough not to accept the "truth" as we see it.

Indeed, seeing is not necessarily believing, and it never has been. During medieval times, the nature of light and the mechanism of vision were both grossly misunderstood. The science of optics was in its infancy, based largely on the work of eleventh-century Egyptian physicist Ibn al-Haitham (Alhazen). In the thirteenth century, Franciscan monks Roger Bacon and John Peckham and Polish physicist Vitellius (Witelo) published visionary treatises on optical instruments and light. Optical theories were often informed by superstition and preconceptions instead of by experiment and objective observation. Within this atmosphere charged by misinformation about sight and by distrust of accessories designed to "help" one see, spectacles were invented. The makers of the first spectacles in the thirteenth century tiptoed around explanations of how their glasses worked; they simply didn't know. That spectacles helped aged eyes see clearly again was sufficient reason for the learned to seek them out.

For the next three centuries, opticians plied their spectacles to the public, unaware that in their hands lay the raw materials for a host of lens-based devices that could be used to investigate nature. It was not until the early 1600s that such optical instruments appeared, the offspring of an unprecedented cross-fertilization between the practical and theoretical realms of study—plus a healthy dose of trial-and-error experimentation. Of these instruments, the most far-reaching, both literally and metaphorically, was the telescope. It was precisely this telescopic reach that would prove necessary to bring stellar parallax within the astronomer's grasp.

Authoritative voices from antiquity, such as Plato's and Ptolemy's, claimed that the eye emits rays that render objects visible. (Think of Superman's "x-ray" vision.) Democritus, among others, held the counter-belief that the eye is a passive instrument and that ordinary objects "sprayed about" visible emissions. Questions abounded. How does a small object generate sufficient emissions for an entire crowd to see simultaneously? How do colors originate? What causes rainbows? How does the image of a large body—a mountain, say—squeeze into the eye? Nobody at the time considered that nonluminous objects are visible because they reflect ambient light to the eye. With vision's underlying mechanism so mysterious, is it any wonder that when Tycho Brahe first saw the supernova of 1572, he asked bystanders to confirm the reality of what he was seeing?

The vista through the typical medieval window pane must have appeared distorted and tinted because of the jumbled consistency and surface irregularities of the glass and the chemical impurities trapped within. A transparent crystal held up to the eye reveals a fractured, rainbow-tinged view of the surroundings. So it's no surprise that shaped pieces of glass—lenses—were viewed by many with suspicion.

The word *lens,* from the Latin for "lentil," is a quaint and overtly nontechnical term that reveals its own artisan origin. Evidently, most medieval scholars weren't much interested in lenses. An exhaustive modern-day archival search turned up only three written accounts concerning spectacle lenses between the years 1280 and 1580. Nobody understood how lenses worked, and during this period it seems that hardly anyone bothered to find out. Many scholars suspected that lenses fundamentally alter the true perception of the world. One thirteenth-century philosopher echoed the prevailing view when he wrote, "The purpose of sight is to know the truth: glass lenses show images larger or smaller than the real ones seen without lenses; they show objects nearer or further away, at times even upside-down or distorted and iridescent; therefore they do not show the truth; hence we must not look through lenses, if we do not want to be deceived."

For the brain to perceive a crisp image, light entering the eye must be concentrated by the eye lens onto the retinal cells that line the rear of the eyeball. The eye lens is convex, that is, thicker in the middle than at the periphery. Were it completely rigid, only objects situated within a certain interval of distance would be perceived clearly. The eye's extraordinary ability to focus light from both nearby and faraway objects stems from the fact that its lens is flexible. The ciliary muscles alter the convexity ("bulginess") of the eye lens in accordance with the distance of the object being viewed: The more convex the lens, the more refractive power it has. For example, to bring into focus a close-up object, the ciliary muscles contract, increasing the curvature of the eye lens. Just the opposite occurs for distant objects. (The cornea and the fluid within the eyeball assist the lens in focusing light onto the retina, but their combined refractive power is constant.)

Around the time I turned forty-five, I noticed a decline in my visual acuity. Not a pleasant milestone for an astronomer. I found that I had to squint while reading. Threading a needle became a challenge. And just when I had gained the means and the courage to plunge into the stock market, the company numbers in the Sunday paper turned into mite-sized hieroglyphics. It is a common affliction of age that the eye loses its ability to focus on nearby objects. The condition is called *presbyopia,* or, colloquially, far-sightedness. Like the rest of the middle-age body parts, the eye lens loses some of its flexibility. It no longer can bulge enough to properly refract light from close-up objects onto the retina; therefore, the image perceived by the brain is blurry. (Technically, the *focal point* of the "weakened" eye lens lies somewhere behind the retina, or equivalently, the *focal length* is too long.) Most children can focus just beyond the tip of the nose, but many middle-aged people have trouble "accommodating" to objects closer than about ten inches. When the eye can no longer focus objects within arm's length, reading and writing become difficult. The working careers of many medieval scholars were hampered by the progressive loss of visual acuity. Indeed, the fourteenth-century Italian poet Petrarch complained about the deterioration of his eyesight when he approached sixty.

Initially, the only way to counter the effects of presbyopia was to place a rounded chunk of quartz or glass onto the manuscript page and magnify the words one by one. This solution was already known to the ancients; indeed, Nero is said to have used an emerald as a magnifier. Magnifying crystals were sometimes embedded in the sides of reliquaries so people could more clearly see the objects inside. According to historical documents, sometime between 1280 and 1290, an Italian craftsman, whose name is lost to history, placed a pair of thin lenses in a wire frame that perched on the bridge of the nose. Dominican friar Alessandro della Spina from Pisa reportedly was shown the first spectacles by the secretive inventor, then promptly reproduced them and made them available to the public. Giordano da Rivalto, another Dominican friar from Pisa, delivered a sermon on February 23, 1305, in which he claims to have met the same inventor twenty years earlier, but fails to name him. A marble tablet in a Florentine church supposedly credited resident lensmaker Salvino degl' Armati with the invention of spectacles, but that claim proved to be a hoax. As historian Vasco Ronchi summed up the situation, "Much has been written, ranging from the valuable to the worthless, about the invention of spectacles; but when it is all summed up, the fact remains that the world has found lenses on its nose without knowing whom to thank." All that is certain is that from this tangled history sprang the precursor of today's multibillion-dollar eyeglass industry.

The first spectacles were nothing more than a pair of magnifying glasses whose handles were riveted together at their ends. Frames were made of wood, metal, bone, or leather. Lenses were

Reproduction of four-teenth-century riveted spectacles.
Source: *Optical Museum Jena, Ernst Abbe Foundation, Germany.*

formed from glass, quartz, or beryl, the latter having a distinct greenish or bluish tint. The lenses were ground manually by rubbing a glass or crystal disk against a shaped dish lined with sand or emery. Later, lathes were introduced to speed up the grinding process. Final polishing was accomplished with a powdery abrasive, such as rottenstone or tripoli.

Early spectacles perched on the bridge of the nose, a precarious situation that led to a host of inventive, if undesirable, ways to better secure them in front of the eyes: hooking them to a hat brim, attaching them to a metal plate strapped to the forehead, tying them with strings around the ears, clamping them onto the temples, threading them into the hair, or forming them into goggles that strapped around the head. Regardless of all the energy lavished on the design and fabrication of spectacles, the final product was often deficient. Even as late as 1770, one spectacle-wearer complained, "They are badly polished, which affects their transparency, there is never the same thickness in the two glasses, their material is usually thready, filled with bubbles and other imperfections."

The development of spectacles (in Italian, *occhiali,* or in Dutch, *brillen*) was a godsend to presbyopic academics of the time. By placing before each eye a simple convex lens, words on the page popped right into focus. A book could be read at a comfortable distance again. Writing was no longer arduous (at least in its mechanical aspects). Today we understand that the convex-shaped glass supplements the weakened refracting power of the eye; spectacle lens and eye lens work together to focus light precisely on the retina. Back in the middle ages, however, the clarifying force of spectacles must have seemed like magic, if not a minor miracle. The suspicion surrounding lens-based vision aids diminished, and spectacles became a symbol of wisdom, even sanctity. In anticipation of the modern-era product endorsement, revered figures began to appear in works of art holding or wearing spectacles. Pythagoras and Virgil; Saints Peter, Paul, Jerome, and Augustine; and, yes, even the infant Christ all found spectacles to their liking. The earliest portrayal of spectacles dates to 1352 in a Treviso fresco panel by Tommaso Barisino of Modena. Pictured at the beginning of this chapter, Hugh of St. Cher Cardinal Ugone is seen poring over a manuscript while wearing riveted spectacles. (The work is pure fiction; Cardinal Ugone died more than twenty years before spectacles entered the scene.) The oldest surviving spectacles, from the late fourteenth century, were discovered during a renovation of a monastery in Wienhausen, Germany.

Spectacle shops opened first in northern Italy, then spread to the Netherlands and from there throughout Europe. By the end of the fourteenth century, one of the fundamental optical components of the astronomical telescope—the convex lens—could be purchased over the counter in many major cities.

The second common ocular ailment, *myopia*, or nearsightedness, occurs when the eye lens is unable to stretch thinly enough to properly focus light from faraway objects (or when the eyeball itself is elongated). In this case, the eye lens is too refractive, too strong. The image it creates does not fall on the retina itself, but somewhere short of the retina. Contrary to the farsighted situation, the focal point is in *front* of the retina and the focal length of the eye lens is too *short*. The myope might be able to read just fine, but distant landscapes are a blur.

By the 1450s, Italian spectacle makers had learned to counter myopia with *concave* lenses, ones that are thinner in the middle than at the periphery. The concave lens slightly disperses light before the light enters the eye, and thereby compensates for the "too-strong" eye lens. (Modern physicists envision a new generation of precision contact lenses and refractive surgeries that might endow the human eye with "supernormal vision.") Concave spectacles appeared in number shortly after the advent of the movable-type printing press and the attendant dissemination of books throughout Europe. Evidently, the learned were transforming themselves into a tribe of myopes; we've since learned that extensive reading, writing, or other close work is one cause of the condition. An early twentieth-century study confirmed a higher frequency of myopia among academics and office clerks than among the non-literate populace. Historian Albert van Helden has even suggested that the frequency of myopia might be used as a rough indicator of literacy in various subpopulations.

By the early 1500s, European spectacle shops stocked a surfeit of convex and concave lenses, all an inch or two in diameter. In keeping with the optical needs of the clientele, the array of convex lenses probably had focal lengths of no more than twenty inches. Lenses of longer focal length refract light too weakly to correct vision substantially. Similarly, the variety of concave lenses in use likely had focal lengths between eight and twelve inches because these were the strengths used to treat myopia. Thanks to efforts to combat natural and induced deterioration of human vision, practically every European spectacle maker in the early 1500s possessed the essential elements to make a crude telescope. So if a telescope is nothing more than two lenses placed one in front of the other within a tube, why did it take the better part of a century for someone to put one together?

There is evidence to suggest that a few spectacle makers and astronomers had considered the idea of making a telescope. But a telescope on paper is one thing; a working telescope quite another. From the wide inventory of lenses, the spectacle maker would have had to select those of appropriate shape and focal length. (A telescope cannot be constructed from just *any* pair of lenses.) The optical performance of a telescope assembled from pre-1600s, "off-the-shelf" spectacle lenses would have been abysmal. The quality of spectacle lenses was generally poor. Lens grinding techniques were primitive and objective testing methods nonexistent. It proved difficult to endow glass disks with precisely curved surfaces and then to polish out all the pits and grooves. And, of course, there was the glass itself.

The fundamental ingredient of simple glass is sand. Glass is basically fused silica, and sand is a convenient source of silica. Chemists know silica by its technical alias, silicon dioxide, after its constituent elements, silicon and oxygen. Silica comes in a variety of forms. The crystalline and amorphous forms are exemplified by the common mineral quartz and by semiprecious opal, respectively. Bastard cousin to both of these is silica's ubiquitous impure form, the kind that wells up between your toes at the beach. So the prime ingredient of glass can be scooped up in your hand, whether at the playground, in the desert, or by the ocean's edge. But to fuse silica into glass requires heat—*plenty* of heat—plus a few trade secrets, whose "unauthorized" disclosure in these pages might have gotten me killed several centuries ago.

Natural glass has been around since the Earth's creation. Obsidian, for example, is a volcanic glass that was used in aboriginal cultures to make cutting tools and arrowheads. Meteors and asteroids sometimes give rise to glass fragments, both during their fiery descent through our planet's atmosphere and when they strike the ground. The heat generated by a lightning bolt is sufficient to transform a chunk of quartz into glass in an instant.

Manufactured glass dates back at least 4,000 years to the Middle East. According to Pliny the Elder's long-ago account, the first such glass formed quite by accident beneath the campfire of some Phoenician traders near the Mediterranean coast. The heat of the fire supposedly melted the underlying sand to yield an opaque, glassy lump. Modern archaeologists have since shifted the likely birthplace of man-made glass to the ancient kingdom of Akkadia in southeast Mesopotamia.

Perhaps the oldest surviving fragment of manufactured glass, dating from around 2000 B.C., was found at an archaeological dig outside the ancient city of Ur, near the confluence of the Tigris and Euphrates rivers. R. H. Hall, leader of the 1918 British Museum expedition, described the artifact as "a lump of opaque blue vitreous paste," which he discovered "in the rubbish beneath the pavement." The oldest dated document about glassmaking is a cuneiform tablet from the seventeenth century B.C., unearthed near the former site of Babylon. From Mesopotamia, glassmaking spread to the eastern Mediterranean coast, where glassblowing was later invented during the first century B.C. Clear glass was developed shortly thereafter in Alexandria. Egyptian campaigns of conquest brought the glassmaker's art to the land of the pharaohs.

Phoenician traders and, later, Roman invaders transported glass products, and eventually the glassmakers themselves, across the Mediterranean to Europe. Glass windows were among the artifacts unearthed at Pompeii and Herculaneum, which were buried in the 79 A.D. eruption of Mt. Vesuvius. In northern Italy, a great glassmaking center arose in Venice. The city was ideally situated: Wood for the furnaces came from the forests of nearby Yugoslavia, sand from the Lido and Verona, clay for the crucibles from Vincenza. Venetian ships brought soda from Egypt. By the end of the sixteenth century, some 3,000 glassmakers were at work in Venice.

The manufacture of glass is a complex process that involves the mixing of disparate ingredients under extraordinary conditions. It is surely the almost magical transformation of "ugly-duckling" materials into a glistening, silky-smooth essence that gave rise to the cult of secrecy surrounding the creation of glass. Secrets of production were tightly held within glassmaking families and artisans' guilds. Governments sometimes took drastic action to protect their own economic interests in the lucrative glass trade. In late thirteenth-century Venice, for example, the city's glassmakers were forced to pack up and shift their operations to the nearby island of Murano. Officials argued that the fires of the glassmaking furnaces endangered the city's residents. In fact, the transfer offshore to Murano may have been orchestrated to remove native glass masters from the temptations and prying eyes of non-Venetian competitors. The penalty for revealing trade secrets to outsiders was death. Venetian officials sent assassins to hunt down fugitive glass masters, and after the deadly deed was done, church bells in Murano pealed in celebration. Even today, secrecy reigns in at least one Venetian glass factory:

> *The room where the raw materials for glass are prepared is partitioned into two sections. Workers in one section place a quantity of each material on a scale, but cannot tell its weight. The scale's readout lies beyond the partition, where the shop's owner yells a full-throated "Basta!" when the scale registers, for his eyes only, the precise numbers called for in the recipe.*
> *It is not surprising to find the work of some Venetian glassmakers—those who experiment with new formulas—jealously guarded, for have they not long been known as* l'uomo di notte, *men of the night?*

Not all sands can be used in the manufacture of glass. Whether from beach, riverbank, or desert, sand typically contains impurities, which from the glassmaker's perspective include any-thing nonsiliceous. Such impurities will tint the glass or even render it opaque. (The oldest surviving glass artifacts are not transparent.) Whenever possible, the ancient glassmakers sought out the relatively rare, fine white sand of quartz-lined river beds. Even into the 1800s, American glass companies offered rewards to those who discovered such sand beds near existing factories. Nowadays, the raw sand is thoroughly washed, then heated to drive off volatile impurities.

Imagine the heat required to melt sand and you'll appreciate how difficult glassmaking must have been for the ancients. The wood- or coal-burning furnaces of antiquity were probably capable of achieving temperatures as high as 2,000 degrees Fahrenheit. This was more than hot enough to smelt metals such as copper, bronze, or iron. Yet even 2,000 degrees is insufficient to render ordinary sand into glass. The discovery that jump-started the glass industry was the observation that certain substances, called *fluxes,* when compounded with sand, create an aggregate whose melting temperature is considerably lower than that of sand alone.

One material that significantly lowers the melting point of silica is *soda* (not the soft drink, but sodium carbonate or sodium oxide). The ancients obtained soda from the remnants of evaporated salt water or by incinerating certain types of salt-marsh plants. Soda alternatives include potash (potassium carbonate), derived from the ashes of hardwood trees; and natron, a mineral soda found in desert deposits that the ancient Egyptians used as a desiccant to preserve mummies. The introduction of soda into the glassmaking recipe reduces the ordinarily high melting point of silica from about 3,000 degrees Fahrenheit to less than 2,000 degrees. The resultant glass is soluble in water; hence its name, "water glass." The further addition of lime (calcium carbonate or calcium oxide), which the long-ago artisans derived from animal bones and shells, strengthens the glass and renders it impervious to moisture. Such soda-lime glass is the form commonly used in windows and bottles. (It is also known as "crown glass," from the old window-making process in which glass was blown into a bub-blelike "crown" before being flattened and cut.) To endow soda-lime glass with color, various elements were added, if they did not already contaminate the raw sand: iron or chromium for green, cobalt for blue, manganese for purple, selenium for red. With rare exceptions, the best "transparent" glass of antiquity would strike us today as distinctly green and cloudy.

It was subsequently learned that tossing in some shards of previously made glass, or *cullet,* catalyzes the melting process. Once melted, this stew of silica, soda, and lime must be "simmered" at high temperature for several days. The lengthy firing is necessary to transform the entire mixture to glass. Otherwise, the result is a *frit:* a jumbled aggregate that is glassy only on the surface. The mixture must be cooled gradually for several days to prevent the solidified glass from cracking.

No one knows the circumstances under which all of these long-ago discoveries were made, nor who made them. Somehow the convoluted process of making glass was pieced together by the ancients over the centuries, despite the many hurdles that had to be surmounted: "the appearance of the ingredients give no hint to the result; the ingredients are not found associated naturally; the right siliceous stone must be employed; an alkaline flux must be added to the pulverized stone; the resultant frit must be refired; a high enough temperature must be attained and maintained for enough time to achieve complete vitrification; and cullet must be introduced to catalyze the process." Only after all this is accomplished, in the proper order and using the proper proportions, does glass result.

In the end, it was this unlikely substance, soda-lime (crown) glass, that medieval spectacle makers inherited from antiquity. Flawed as it was, this was also the glass the spectacle makers would eventually fashion into the first primitive device to view the heavens.

Peer through a window of a medieval-era building, and you'll realize why the common glass of the day was unsuitable for lenses in optical instruments: It was inhomogeneous, it contained trapped air bubbles and streaks ("striae"), and it was not fully transparent. The production of higher-quality glass required the finest white sand or quartz and special procedures in the firing and cooling processes.

Pre-Renaissance scholars showed little interest in promoting the development of high-grade optical instruments, much less in putting them to scientific use. Whatever they needed to know, in their opinion, could be found in the writings of the ancients. Their curiosity about nature was satiated by the judgments—and misjudgments—of antiquity, as opposed to direct observation. Without any call for optical devices more sophisticated than spectacles, market forces during the 1500s dictated against improvement in lens-making materials and methods and against experimentation with lenses. However, the situation was about to change with surprising swiftness.

Shortly after 1600, several spectacle makers awakened almost simultaneously to the notion that a crude, low-magnification "spyglass" might have commercial value. (This development is a prime example of what writer Malcolm Gladwell describes as a "tipping point" in the societal spread of an idea.) The spectacle makers envisioned their spyglass as a military surveillance tool to spy on the enemy, or even as an irresistible novelty item, the Renaissance-era predecessor of that popular nineteenth-century optical toy, the kaleidoscope.

I can picture one of these long-past spyglass inventor-entrepreneurs sitting, as I am now, in front of the window. Beside him is a tray like the one I've placed on my writing table, containing a jumble of "Renaissance-era" lenses. We are both aware of how a single lens affects one's vision: He makes spectacles for a living, I teach optics to my college students. But what will we see when we look out the window through various *pairs* of lenses?

Sorting through our old-style lens collection, we select samples from three basic categories:

- Strong convex lenses (focal length of, say, six inches)
- Weak convex lenses (focal length about twenty inches)
- Concave lenses (our entire concave inventory falls within the narrow focal-length range of eight to twelve inches, so we lump them all into one category)

These three lens types are sufficient to paint a relatively complete picture of the behavior of various lens pairs. From these types, we can establish *nine* different pairs to test. (Looks like it may not have been so simple to invent the telescope after all.) The plan is to hold one lens before the eye—the "eyepiece"—and place in front of it the second lens—technically known as the "objective" lens. Then we'll observe what we see when the two lenses are separated. The following table lists the nine possible configurations:

	Objective Lens	Eyepiece Lens
1.	strong convex	strong convex
2.	weak convex	weak convex
3.	strong convex	weak convex
4.	weak convex	strong convex
5.	concave	concave
6.	concave	strong convex
7.	strong convex	concave
8.	concave	weak convex
9.	weak convex	concave

Now it's time to experiment. Let's begin with configuration 1, a pair of strong convex lenses. I hold before my eye a single convex lens (the eyepiece lens) and look out the window at my neighbor's house. The house appears its normal size, although blurry. Holding the other convex lens (the objective lens) close to the first, the house looks even blurrier. The situation gets worse as I move the objective lens away from the eyepiece lens. When the two lenses are separated by a few inches, the image of the house loses its form entirely. But wait. When I move the main lens farther yet, the house reappears, slightly magnified, still somewhat blurry—and now upside-down. Separating the lenses more merely makes the house blurry again. So we arrive at a counterintuitive result that two strong magnifiers, when placed in tandem, fail to produce a usable magnifying system. A strong convex/strong convex lens pair would not have impressed our would-be spyglass inventor.

Running down our list, the next two configurations—weak convex/weak convex and strong convex/weak convex—turn out to be equally unimpressive. However, the fourth configuration, a weak convex objective paired with a strong convex eyepiece, looks promising. When I place the two glasses against one another in front of my eye, my neighbor's house appears upright and slightly magnified, but very blurry. Moving the weak objective lens away from the strong eyepiece lens doesn't improve matters until the objective is way out at arm's length. Now my neighbor's house appears flipped-over, quite distinct and magnified about threefold.

A telescope? Definitely. Practical? Were I in the shoes of our long-ago spectacle maker, I would answer no. The device is uncomfortably long, well over two feet, given the lens pair that was selected. Keeping such far-apart lenses properly aligned might prove troublesome. Also the image remains clear only when the lenses are separated by a precise amount, another hassle from the production standpoint. And finally, who'd pay good money to see the world upside-down? Time to reach back into the lens tray. (In fact, it was this design that ultimately triumphed and became the modern refractor telescope. Stunning magnifications were achieved once lens makers learned how to produce both very weak convex objectives and very strong convex eyepieces. Astronomers didn't mind seeing the universe upside-down.)

Dipping now into our concave-lens supply, a concave/concave combination (configuration 5) merely shrinks the world, no matter how the lenses are arranged before the eye. So does configuration 6, a concave objective coupled with a strong convex eyepiece. Configuration 7, with a strong convex objective and a concave eyepiece, does magnify when the lenses are not too far apart, although the view of my neighbor's house is always blurry; separating the lens pair more clears up the image but destroys the magnification. Similarly, configuration 8, a concave and a weak convex, yields no telescopic effect.

For configuration 9, I hold the concave lens up to my eye and place the weak convex objective over it. My neighbor's house looks sharply defined and upright, but not magnified. I move the convex objective lens farther away. Eureka! The image of the house gets larger, yet remains upright and distinct. The lenses are now about a foot apart. The house appears approximately threefold magnified. And clear. Separating the lenses

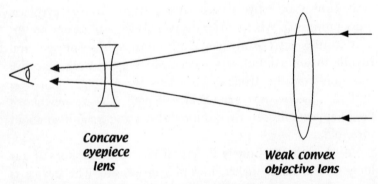

**Concave
eyepiece
lens**

**Weak convex
objective lens**

Convex and concave lenses arranged in configuration 9.

more just makes the image blurry. Here, finally, is the configuration the Renaissance spectacle makers were looking for: a weak convex objective coupled with a stronger concave eyepiece, spaced about a foot apart. A compact spyglass, showing an upright, modestly magnified view of the world. Not precisely what Roger Bacon had in mind when he wrote in the mid-thirteenth century: "Thus from an incredible distance we might read the smallest letters and number grains of dust and sand. . . . In this way a child might appear a giant, and a man a mountain. . . . So also we might cause the sun, moon, and stars in appearance to descend here below."

No, the crude spyglass fell far short of Roger Bacon's imagined instrument. It was a telescope to the same extent that a pair of binoculars is a telescope: The optical principle is identical; only the magnifying power is missing. Nevertheless, the would-be spyglass makers believed they were onto something lucrative. And the first one to market would gain the advantage.

On September 25, 1608, spectacle maker Hans Lipperhey of Middelburg in the Netherlands submitted a patent application to the States-General, the governing body of the Dutch federation, for "a certain device by means of which all things at a very large distance can be seen as if they were nearby." Here in this document is the first unequivocal description of a practical, telescopelike instrument. (There is an unconfirmed story that Lipperhey's children stumbled upon the magnifying properties of the dual-lens combination while playing in their father's workshop.) Lipperhey's spyglass stirred considerable interest. An examining committee hastily convened. Indeed, as claimed, faraway objects looked closer through Lipperhey's device. In all likelihood, this Dutch telescope-wannabe exhibited all the flaws of a modern, ninety-nine-cent toy spyglass: indistinctness, distortion near the periphery of the view field, rainbow halos around the images. But to Renaissance-era sensibilities, the sheer reach of Lipperhey's instrument was unlike anything that had been seen before.

The committee made two recommendations: first, that Lipperhey submit an upgraded spyglass with lenses of quartz instead of glass; and second, that the new instrument be binocular, so the user won't have to squint with one eye closed. Despite Lipperhey's desperate plea for secrecy, a diplomatic newsletter from the Hague circulated throughout Europe, describing the spyglass and stating that "the said glasses are very useful in sieges and similar occasions, for from a league and more away one can notice things as clearly as if they were quite near us; and even stars that ordinarily do not appear to our sight and our eyes because of their smallness [faintness] and the weakness of our sight can be seen by means of this instrument."

On October 14, 1608, just three weeks after Lipperhey submitted his application, a second patent request for a spyglass arrived before the committee. This one came from a fellow townsman of Lipperhey's, twenty-year-old Sacharias Janssen. Spectacle maker, traveling merchant, and convicted counterfeiter, Janssen claimed to have invented the spyglass years before Lipperhey. (Janssen's son later testified under oath that Lipperhey had stolen the idea from his father in 1590. Apparently, the son had neglected to do the math: In 1590, Sacharias Janssen was only two years old.) One day later, on October 15, 1608, Jacob Metius, a well-regarded spectacle maker in Alkmaar in northern Holland, presented yet a third patent application for a spyglass. During this flurry of competing claims, word arrived that a Dutch merchant had exhibited a spyglass at the annual autumn fair in Frankfurt.

The Dutch government concluded that the spyglass was not patentable. It was simply too easy to make! Lipperhey was amply compensated for the three binoculars he ultimately produced for the state. Subsequent testimony suggested that the Dutch spyglasses were, in fact, not original, but copied from a 1590 specimen imported from Italy. Indeed, Middelburg's vaunted glass foundry, established in 1581, was managed by an Italian and had several Italian employees, any one of whom might have constructed a spyglass or received one from back home. Unfortunately, Middelburg's city hall, whose archives held a wealth of documents relating to this issue, was destroyed by the Luftwaffe during World War II.

The spyglass took Europe by storm. By April 1609, spyglasses were on sale in Paris, by May in Milan, and by August in Naples and Venice. As the makers had foreseen, nations adopted the instrument as a high-tech "early warning system" to detect enemy troops or ships. The general public flocked to this must-have novelty, this magical deflater of distance. Only the scientists were underwhelmed. They saw the spyglass as a curiosity, neither powerful enough nor trustworthy enough to be employed in research. Even the astronomers didn't clamor for them. Sure, the spyglass would make heavenly bodies appear closer. Still, what sense would it make to train the device on the night sky? What could it possibly show that they, the astronomers, didn't know already?

Galileo Galilei at age forty-two, by Domenico Robusti (Tintoretto).
Source: *Copyright, National Maritime Museum, London.*

7

The Wrangler of Pisa

Who shall teach thee, unless it be thine own eyes?
—*Euripides,* Helen

The Bible tells us how to go to heaven, not how the heavens go.
—*Cardinal Baronius, 1615, often attributed to Galileo*

Galileo Galilei, professor of mathematics at the University of Padua, was having trouble making ends meet. It was 1609, now eighteen years since his father Vincenzio had died. In the intervening period, Galileo had become the de facto head of an extended family. He still supported his mother and, at various times, his three younger siblings. (Three others had died in childhood.) He had willingly shouldered the onerous dowry arrangements that had been contracted for his sisters, Virginia and Livia; the down payment for Livia's dowry alone had been twice his annual salary at Padua. Nor had he spared any expense on Livia's wedding or on the celebratory feast that followed. He had long ago written off the loan he had given his musician brother, Michelangelo, to travel to Poland; only last year, Michelangelo was married in a gala affair—courtesy of Galileo. And Galileo himself, now forty-five years old, had three children of his own to support, along with his common-law wife, Marina Gamba, all of whom lived in a separate house on the Ponte Corvo, a few minutes' walk from where Galileo lived.

Mathematics professors in early 1600s Italy typically made about one-sixth the salary of their counterparts in more valued fields such as philosophy or medicine. To supplement his university pay, Galileo ran a home business where he produced precision drafting and measuring instruments. With his invention of a calculating device he called a "geometric and military compass," business was good. Like many of his colleagues, Galileo earned additional money tutoring students and renting out rooms in his house. Still, he'd been forced to request advances on his salary and once even to borrow money from his friend Giovanfrancesco Sagredo, a Venetian gentleman and amateur scientist.

Life, in short, was a rat race for Galileo Galilei. He longed for the chance to rid himself of the financial pressures and devote his energies entirely to his research on the physics of motion—if only the right opportunity would come along.

Galileo might have sprung straight from the brow of his father, Vincenzio, so alike were the two in their natures. Both displayed a purposeful dichotomy of behavior that marked their respective paths through life: sometimes generous, sometimes self-interested; sometimes straightforward, sometimes evasive; sometimes outspoken, sometimes circumspect. In terms of career, Vincenzio Galilei was a down-on-his-luck cloth merchant; in terms of avocation, he was a passionate and opinionated patrician who wrote treatises on music theory and stormed the edifice of convention. In Vincenzio's manifesto, *Dialogue on Ancient and Modern Music,* is written this frank statement: "It appears to me that they who in proof of anything rely simply on the weight of authority, without adducing any argument in support of it, act very absurdly. I, on the contrary, wish to be allowed to raise questions freely . . . as becomes those who are truly in search of the truth." Vincenzio's declaration could well have served as the eventual watchword of his son Galileo. Study the father and you'll learn where the son got his dogged personality, polemical flair, and relentless impulse to question both nature and authority.

As a youth, Galileo showed talent for art, lute-playing, and stubbornness. He was educated at the Benedictine monastery in Vallombrosa. However, when the adolescent Galileo expressed an interest in becoming a monk, Vincenzio yanked him out of the monastery and brought him home. By 1581, Vincenzio had scraped together enough money to enroll Galileo, then seventeen, at the University of Pisa to study medicine. Medical instruction at Pisa, as at every other European university, rested on teachings of the ancients: the anatomy of Galen; the philosophy and science of Aristotle; even the astronomy of Ptolemy, for the human body was thought to be influenced by the cycles of the planets and stars.

Galileo was unimpressed with the explanations of the ancients; he openly questioned their wisdom, relying instead upon his own powers of observation. He contradicted his professors and, when faced with intellectual rigidity, often responded with sarcasm or even outright hostility. One observer described Galileo's favorite polemical technique: "[B]efore answering the opposing reasons, he amplified them and fortified them himself with new grounds which appeared invincible, so that, in demolishing them subsequently, he made his opponents look all the more ridiculous." Even in private, Galileo could not restrain his annoyance when confronted with stupidity. In the marginal notations of an erroneous paper on comets, he savages the author in colorful Italian: *ridiculoso, elefantissimo, ingratissimo villano, villan poltrone*—ridiculous, elephantine, ungrateful bumpkin, lazy bum. It's easy to picture Galileo shouting to the heavens about the mediocrity around him. The Pisan faculty, annoyed and angry, adopted their own name for this assertive son-of-a-cloth-merchant: "The Wrangler."

To his father's dismay, Galileo's medical studies fell by the wayside when he switched his attention to mathematics and physics. With the help of family friend Ostilio Ricci, mathematician to the Duke of Tuscany, Galileo plowed through Euclid's *Elements*. He left the University of Pisa in 1585 a skilled mathematician, but still several courses shy of a degree. The next four years were spent performing physics experiments and foraging for work. Galileo tutored mathematics and delivered lectures, including an invited address before the Florentine Academy on the location and layout of Hell, as culled from Dante's *Inferno*. Galileo also spent plenty of time forging connections within the intellectual and political ranks, that is, "schmoozing" with important people.

In the view of biographer Stillman Drake, "The picture of Galileo's personality presented year by year in his correspondence and publications is that of a prudent man, not given to forming conclusions without having weighed the evidence, well aware of social customs, and disinclined to quarrel with highly placed persons in Church or state." For now, Galileo kept his controversial ideas close to his vest, publishing no academic treatises and circulating his writings only in manuscript form among friends. He was a pious man and had no desire to be regarded either as prophet or revolutionary. The aesthetics of the Copernican universe appealed to him, yet he publicly taught and defended the reigning system of Ptolemy. Although outspoken, Galileo understood that career advancement would not come from battering too noisily at the gates of the establishment.

In 1589, Galileo drew upon his political contacts, and not for the last time. To the chagrin of the Pisan faculty, Galileo was hired as a mathematics professor at his alma mater. He viewed his low-paying appointment at Pisa as a stepping-stone to a more lucrative and prestigious post. In the meantime, he relied on his influential friends to pursue his cause. He settled into a routine of teaching, research, and "wrangling" with the more conservative of his colleagues. That he refused to wear the standard-issue academic robes only antagonized his opponents more. From the start, a core of Pisan professors worked to ensure that Galileo's three-year contract would not be renewed.

While Giacomo della Porta, one-time assistant to Michelangelo, was putting the finishing touches on the great dome of St. Peter's, Galileo in Pisa was making his initial discoveries about natural motion. First, he developed the physical basis of the pendulum clock. He determined that the time it takes a pendulum to complete a swing is the same regardless of how wide the arc of the swing. As to why a pendulum should take the same time to complete a large swing as a small one, Galileo realized that a large-swinging pendulum's faster speed precisely negates the effect of its longer travel path. That's why a grandfather clock ticks at a constant rate as its pendulum "winds down." (Actually, Galileo's observation about the pendulum is true only when the pendulum's swing is not too large.) According to an account by one of his former students, Galileo's insight sprang from the oscillation of a particular candelabra in the Cathedral at Pisa. An inspirational tale—except the candelabra in question was installed years after Galileo recorded the idea.

Galileo also disputed Aristotle's assertion that heavier bodies fall to Earth faster than lighter ones. Galileo had already observed that hailstones arrive at the ground in a mixture of sizes; were Aristotle right, larger hailstones would reach the ground first, followed by progressively smaller ones. Galileo supposedly dropped a pair of unequal-weight cannonballs from the Tower of Pisa to prove that they strike the ground simultaneously. In all likelihood, Galileo carried out such an experiment, but not from the famous Tower. He concluded that Aristotle was wrong; all bodies fall to Earth at the same rate, regardless of weight. In fact, the heavier cannonball would have struck the ground slightly ahead of the lighter one because of the unequal influence of air resistance on the two cannonballs. Galileo understood this quite well. The identical falling rates of objects can be perfectly demonstrated only in a vacuum. Every year in my astronomy course, I show the Apollo-era filmstrip in which an astronaut on the Moon releases a falcon feather and a hammer simultaneously from chest height. The feather plummets just as fast as the hammer, and both strike the lunar surface at the same instant.

In 1592, Galileo's Pisan adversaries got their wish. The Wrangler, having acquired some measure of acclaim for his ideas and inventions (and having made further use of his political contacts), accepted a higher-paying post at the University of Padua, where Copernicus had studied medicine a century earlier. Some twenty miles from Venice, Padua was more liberal and enlightened than the Tuscan city of Pisa, having benefited from six generations of tolerant Venetian rule. The intellectual environment extended beyond the university's walls. In Padua, Galileo found many like-minded scholars who supported his investigations into the nature of motion, if not always the conclusions. It was here, too, that Galileo met his companion, Marina Gamba, and started a family.

In August 1597, Galileo received a book entitled *Mysterium Cosmographicum,* "Mystery of the Cosmos," written by a young German mathematician named Johannes Kepler. Replete with mystical ramblings and numerology, *Mysterium Cosmographicum* is nonetheless a powerful meditation on the harmony of the Copernican system. In his letter of thanks, Galileo mentioned to Kepler that he had "adopted the teaching of Copernicus many years ago," but had not declared himself publicly for fear of professional ridicule. Kepler was obviously pleased to hear from this literate stranger whom he initially described to a friend as the "Italian whose last name was the same as his first." He replied on October 13, 1597, urging Galileo to go public in support of the Copernican system: "Stand forth, O Galileo!"

In the same letter, Kepler asks Galileo whether he has any astronomical instruments capable of measuring star positions with an accuracy approaching a thousandth of a degree. With such a device, Kepler declared, Galileo might detect the parallaxes of stars and thereby confirm the Earth's orbital motion. "Even if we could detect no displacement at all, we would nevertheless share the laurels of having investigated a most noble problem which nobody has attacked before us." Galileo didn't share the enthusiasm of his German counterpart, whose verbal bravado marked him as a trigger-happy Copernican zealot (which he was). In any event, Galileo was powerless to do anything about the parallax issue. He wasn't an observational astronomer. He had no instrument of the precision Kepler suggested and had no intention of making one. Twelve years would pass before Galileo wrote to Kepler again, with news so stunning that it changed the course of science. In the meantime, Kepler sought out the one astronomer who did have instruments of the requisite precision: Tycho Brahe.

As the year 1609 neared its midpoint, Galileo Galilei could look back on the previous decade at Padua as a productive one. True, he'd been saddled with money worries and the ever-present interruptions of his research. Nevertheless, he had amply provided for his family and had made quite a reputation for himself as a precision-instrument maker and an expert in the mechanics of motion. It had been a decade of novel experiments involving inclined planes and rolling balls, from which he had established fundamental relationships between distance, speed, and acceleration. Only once during that period had Galileo's scientific attentions strayed into the astronomical arena.

In October 1604, a new star had appeared in the heavens and gradually faded away, just as in 1572. Like its predecessor, the brilliant intruder of 1604 displayed no parallax to observers across Europe; it was truly a remote star. Galileo delivered three lectures in which he openly criticized the old Aristotelian belief that nothing in the stellar realm ever changed. The result was a public feud between Galileo and Cesare Cremonini, senior professor of philosophy at Padua. But the feud had settled down and Galileo had returned to his former Earthbound pursuits. Now, as always, Galileo kept his eyes open, not just to the wonders of Nature, but to any possibility of bettering his situation.

Sometime during the spring of 1609, Galileo's longtime friend, Paolo Sarpi, statesman and official theologian to the Venetian senate, related to Galileo a rumor that had been circulating within diplomatic circles: An unknown Dutch spectacle maker had supposedly invented an optical instrument that made faraway objects appear as though nearby. Galileo had certainly heard rumors of this sort before. The age was rife with apocryphal reports of magical devices, absurd phenomena, and strange beasts. Virtually every one proved to be exaggerated or an outright hoax. Galileo probably thought that this "spyglass" of Sarpi's was no different.

Around June 1609, however, Paolo Sarpi delivered stunning news: A former student of Galileo's named Jacques Badovere had seen a spyglass on sale in the window of a Paris optical shop. Sarpi later reported to Galileo that a foreign merchant had appeared before the Venetian senate, offering—at a substantial price—a crude spyglass for military purposes. Such a device was of great interest to the Venetians. Venice had no defensive wall, but relied instead on its navy to defend the city against sea-based attack; if enemy warships could be spotted sooner, the city would have more time to mount an aggressive defense. Sarpi had advised the senate against the purchase of the foreigner's spyglass. He trusted that his old friend Galileo could make a better one—and reap the financial benefit.

Galileo must have seen the door of opportunity swing wide, having been opened by a most unlikely key: a simple optical device, the spyglass. If the Venetian senate would consider buying an inferior instrument from a stranger, surely they would accept an improved one from their esteemed citizen-scientist. Galileo was in an ideal position at the ideal time; he possessed the optical know-how, the manual skill, and the tools to construct a spyglass of his own. But he would have to work quickly. If one spyglass merchant had found his way to Venice, others would surely follow.

Using his knowledge of optics, Galileo deduced the design of the Dutch spyglass: a weak convex objective lens combined with a strong concave eyepiece. He bought the requisite glasses from a local spectacle maker and mounted them in a lead tube. From design to completion had taken him all of a day. The instrument magnified only slightly, as Galileo had expected. No, he could not tout the merits of such a feeble device before the Venetian senators. To leverage the maximum reward from the senate demanded something unprecedented and unexpected, an instrument far superior to those being hawked by common peddlers. Galileo knew that a spyglass made from ordinary spectacle lenses would never magnify more than threefold. But were he to grind and polish lenses of his own design, using the finest-grade, clear Venetian glass, he might produce an exquisite instrument of double, or even triple, the magnification of any other on the continent. Galileo must have understood quite clearly that in the power of his spyglass lay the power to broker his future.

After numerous trials grinding and polishing glass disks against a rotating abrasive wheel, Galileo found the key to increasing the magnification of the Dutch-style spyglass. First, he had to endow the convex objective lens with the gentlest possible curvature; that is, he had to make its focal length long. Next, he had to grind a relatively deep curvature into the concave eyepiece, making its focal length short. The more disparate these focal lengths, the higher the magnification. Once he had completed an acceptable nine-power spyglass, Galileo asked Sarpi to make the arrangements. On August 25, 1609, Galileo led the Venetian senate up the Tower of San Marco to demonstrate his instrument. "There have been numerous gentlemen and senators," Galileo wrote afterward, "who, though old, have more than once climbed the stairs of the highest bell towers of Venice to observe at sea sails and vessels so far away that, coming under full sail to port, two hours and more were required before they could be seen without my spyglass."

Galileo presented his instrument as a gift to the city of Venice. No doubt impressed with the spyglass and with Galileo's (calculated) generosity, the senators doubled Galileo's salary and offered him lifetime tenure at Padua. Only later did Galileo learn the details: The raise wouldn't kick in for another year, there would be no future raises, and his teaching load would not be reduced. Believing he could do better, Galileo traveled briefly to Florence in his native Tuscany. There he presented a similar spyglass to Tuscany's recently elevated ruler, Grand Duke Cosimo II de' Medici, to whom Galileo had once served as private tutor. Galileo tactfully suggested that Cosimo might hire him as court philosopher and mathematician. With no teaching duties to distract him, Galileo probably pointed out, all of his time could be spent making discoveries to the everlasting glory of the Medici family. While Cosimo deliberated, Galileo returned to Padua and recast his workshop into a veritable spyglass factory. By November 1609, he had completed a twenty-power instrument and, two months later, one of thirty power that was just over an inch in aperture. These were the instruments that Galileo turned to the heavens. There he discovered celestial wonders that had never been seen by anyone since the dawn of the world.

If you've ever tried to hold steady a pair of binoculars, you'll appreciate the challenge Galileo faced when applying his thirty-power spyglass to the night sky. The magnification of celestial light in an optical instrument is always accompanied by its nettlesome sibling: the "magnification" of earthly vibrations. Every twitch of the hand, every tread of the foot, every intake of breath wreaks havoc on the clarity of the image in the eyepiece. The vista

Twenty-one power telescope by Galileo.
Source: *Istituto e Museo di Storia della Scienza di Firenze.*

through this glassy peephole jitters incessantly. The straightforward task of seeing becomes as difficult as trying to watch TV from a lurching carnival ride. Amateur astronomers routinely hold their breath until they turn purple or try not to shiver in the frigid evening air, just to keep themselves and their instruments steady. Watch the astronomer intent upon the eyepiece and you'll witness the creation of a living statue, as a curious sort of in vivo "rigor mortis" settles over the body. To steady his own view of the heavens, Galileo strapped his spyglass to a rigid stand that could be placed on a table. He touched the instrument only when bringing a celestial object into view.

Every astronomer knows, too, that a telescope's eyepiece fogs from moisture in the breath. The fogging seems to occur especially during moments of excitement, when a pulse-pounding view—the Moon's face, a swirling nebula—pours into the eye. Fogged-up eyepieces dogged Galileo, which is not surprising given the celestial sights he saw.

Galileo suffered a further challenge from his spyglass: narrow field of view. Just how restricted was Galileo's view of the night sky? Consider the standard, eleven-inch-long cardboard tube that sits at the center of a roll of paper towels. Shut one eye and peer through the tube with your other eye. The tube masks all but an 8-degree-wide circular "window" on your surroundings. Now imagine a similarly restricted view of the world, this time seen through a *twenty-nine-foot-long* version of that same tube. Such was the extent of tunnel vision afforded Galileo by his most powerful spyglasses, whose field of view was so narrow that only one-quarter of the Moon could be studied at one time. Certainly nobody would buy an instrument like Galileo's today; by modern standards, Galileo's spyglass had but modest magnification, coupled with a chimneylike aspect on the sky. Sighting a particular star or planet through such a spyglass must have been a chore. And keeping the object within the field of view must have been equally difficult. (Celestial objects drift out of the field because of the Earth's rotation.) So when you read below of Galileo's discoveries, remember that the "mere" act of pointing a spyglass to the sky is more complex than it sounds.

On November 30, 1609, outside his home in Padua, Galileo peered through his twenty-power spyglass at the Moon. He must have seen immediately that the Moon was nothing like the smooth, unblemished celestial body postulated by the ancients. The lunar surface, he later wrote, was "uneven, rough, and crowded with depressions and bulges." The spyglass also revealed craters, valleys, and jagged mountains. Some of the mountains were so tall that, when the Moon itself was at crescent or half, the tops of the mountains caught the Sun's rays even as their bases languished in darkness. From the lengths of the shadows cast by these peaks, Galileo determined that the tallest rose four miles above the surrounding plains. There was nothing even remotely "heavenly" about the twenty-times enlarged Moon, nothing that fundamentally distinguished it from our own world. One craterous landscape even reminded Galileo of Bohemia. "It is like the face of the Earth itself," Galileo wrote. And he was right: Looking at the Moon, he saw what appeared to be just another planet.

Next, Galileo trained his spyglass on the stars. To the familiar stars of antiquity, Galileo now added myriad others invisible to the naked eye. In the Pleiades star cluster alone, where the likes of Hipparchus and Ptolemy had tallied just seven stars, Galileo counted forty-three. "To whatever region of [the sky] you direct your spyglass," Galileo declared, "an immense number of stars immediately offer themselves to view." The Milky Way's ghostly glow, a subject of dispute for thousands of years, Galileo now saw for what it was: the collective light of a vast assemblage of stars. And unlike the Moon, whose image grew larger as the magnification was increased, the stars always remained mere points of light, even at the highest magnification. The stars, as Copernicus had suspected, must indeed be extremely far away to render the magnifying power of the spyglass ineffective.

In terms of sheer impact, even the extraordinary observations of the Moon and the Milky Way couldn't measure up to Galileo's premiere discovery: the moons of Jupiter. Viewing the planet from January 7 through March 2, 1610, Galileo noticed a curious line of four "stars" flanking the image of Jupiter. From night to night, these attendant points of light changed position, yet never strayed far from the planet and never relinquished their linear arrangement. Galileo realized that these were not stars in the ordinary sense but previously unseen Moonlike bodies in orbit around Jupiter. No longer was the Earth-Moon system unique. To Galileo, the image of Jupiter accompanied by a phalanx of smaller bodies resembled a reduced-scale version of the Copernican solar system.

Every year in my sophomore astronomy class, we set our computer planetarium program to display the positions of Jupiter and its moons as they had appeared from Padua an hour after sunset on January 7, 1610, the night Galileo discovered them. Sure enough, the computer screen shows the precise arrangement that Galileo had recorded in his observing log on that date. Staring at the tiny, pixelated image of Jupiter and its moons always stirs a parallel image in my mind: that of a long-ago hand steadily, carefully sketching planet and moons on paper with a quill pen. In subsequent observations of Jupiter, Galileo unknowingly observed Neptune. The solar system's eighth planet is marked as a star near Jupiter in Galileo's journal entries for both December 28, 1612, and January 28, 1613, well over 200 years before the "official" discovery of Neptune by German astronomer Johann Galle in 1846.

In March 1610, Galileo released a short book entitled *Sidereus Nuncius*, "The Starry Messenger." (Historian Edward Rosen suggests that the title translation "Message from the Stars" might be more in accordance with what Galileo had intended.) Here, in concise language, Galileo sets forth his celestial observations of the previous months. His intention is evident right from the book's opening line: "In this short treatise I propose great things for inspection and contemplation by every explorer of Nature." *Sidereus Nuncius* wasn't a scientific treatise for academics; it was Galileo's invitation to the literate class: Visit the new universe, *your* universe, not the one dictated by age-old, untrustworthy sages. Throughout the book, Galileo's enthusiasm bursts forth in phrases like "beautiful and delightful sight," "natural excellence," and "absolute novelty." He also implicitly stands with Copernicus, writing that Jupiter and its moons "complete their great revolutions every twelve years about the center of the world, that is, about the Sun itself."

Galileo shrewdly dedicated *Sidereus Nuncius* to Cosimo II, Duke of Tuscany, in the hope that Cosimo would become his eventual benefactor. And in case Cosimo didn't get the message, Galileo christened the newly discovered worlds around Jupiter the "Medicean Stars," after Cosimo and his three Medici brothers: "Behold, therefore, four stars reserved for your illustrious name, and not of the common sort and multitude of the less notable fixed stars, but of the illustrious order of wandering stars, which, indeed, make their journeys and orbits with a marvelous speed around the star of Jupiter." Kepler subsequently suggested in a letter to Galileo that, to avoid confusion, Jupiter's "stars" be referred to henceforth as *satellites,* from the Greek *satellos,* meaning "attendant."

Along with the bound copy of *Sidereus Nuncius,* Galileo sent Cosimo the actual spyglass he had used to discover Jupiter's satellites. Cosimo responded precisely as Galileo had hoped: He named Galileo official mathematician and philosopher to the Tuscan court, with a substantial salary. (The appellation "philosopher" added a cachet of respectability that "mathematician" alone failed to convey.) The Duke also appointed Galileo chief mathematician at the University of Pisa, but specified that Galileo was to be permanently excused from teaching. In fact, Galileo never had to set foot on campus if he so chose. In September 1610, Galileo left Padua and moved to Florence, back to his beloved Tuscany.

The famous spyglass Cosimo had received from Galileo was eventually left to Cosimo's eldest son and heir, Ferdinando II. Over the centuries, the lens from the instrument passed from the Medici family through various galleries and museums. The lens, with a crack running its width, now resides in the Florentine Museum of the History of Science.

Sidereus Nuncius brought Galileo worldwide acclaim. According to a letter Galileo sent to Cosimo, the first printing of 550 copies sold out within a week. Jesuit missionaries carried the book as far as China. Many critics who refused to believe Galileo's observations were silenced in late 1610 when Jesuit astronomers in Rome affirmed the remarkable discoveries with their own spyglass. In response to a plea from Galileo, Kepler published a widely circulated letter of support, entitled "Discussion with the Starry Messenger." By the end of 1610, the tide of acceptance had swung mightily in Galileo's favor. Visiting Rome in the spring of 1611, Galileo was granted an audience with Pope Paul V, who conveyed his respect by refusing to let Galileo remain kneeling. The Jesuit astronomers celebrated Galileo, and clerics lined up to see the heavens through one of his instruments.

On April 14, Prince Federico Cesi and his scientific society, Accademia dei Lincei (the Academy of the Lynxes), held a banquet in Galileo's honor on a hillside just outside Rome. After dinner, Galileo thrilled the assembled guests by pointing one of his spyglasses at the moons of Jupiter and the stars of the Milky Way. When he set his device on the Lateran Church over a mile distant, guests were amazed to see clearly the inscription on the building's façade. The host, Federico Cesi, announced that Giovanni Demisiani, mathematician to Cardinal Gonzaga, had devised a name for Galileo's spyglass, one that more aptly conveyed the instrument's capabilities and differentiated it from its crude, low-power cousins. The name, Cesi explained, was a melding of two words in Demisiani's native Greek: *tele,* meaning "far away," and *skopéo,* "to look." Henceforth, Galileo's instrument would be known as the *telescope.*

Galileo continued his telescopic observations from his new home in Florence. Here he discovered that Saturn sports a pair of curious appendages, but his telescope was not powerful enough to reveal their true nature. (They were Saturn's rings.) He continued his surveillance of Jupiter's moons in the hope of measuring their orbital periods. He observed sunspots and concluded that they were "blemishes" on the Sun's surface, not uncharted planets seen in silhouette. However, it was his observations of Venus that made the biggest impact on the course of his professional life. Galileo found that Venus cycles through phases just like our Moon. As the months passed, Venus's appearance in the eyepiece changed gradually from an almost fully illuminated disk to a crescent, before widening back into a disk again. Simultaneously, Venus appeared to change size: largest when crescent, smallest when full. Whatever Copernican sympathies Galileo might have harbored up to this point became fully realized in the wake of this discovery. In a Ptolemaic universe, it is impossible for Venus to display the complete range of phases; since Venus always lies between the central Earth and the illuminating Sun, a "full" Venus cannot occur. Only the Copernican and Tychonic systems give rise to the entire range of phases; in either case, the full phase occurs when Venus circles the far side of the Sun opposite the Earth. Galileo had long ago dismissed Tycho's model as illogical. The Copernican system was the only viable option.

From Galileo's subsequent writings and orations, it is clear that his allegiance to Copernicus was now complete and unalterable. He had finally "stood forth," as Kepler had urged more than a decade earlier. With missionary zeal, he set about trying to convince the Church itself of the Copernican reality. Galileo believed it was only a matter of time before telescope-wielding astronomers detected the telltale sign of Earthly motion: stellar parallax. And whenever that day arrived, he wanted the Church—his Church, after all—to be on the right side of the world-system conflict. In the early seventeenth century, conservative elements within the Church were gaining strength. The Protestant Reformation of the 1500s had provoked a Counter-Reformation within the Catholic Church, which effectively removed the interpretation of Holy Scripture from lay hands. Galileo feared that the Church, if sufficiently pressured from within, might take the fateful step of declaring the Copernican system heretical—a step that would require an embarrassing reversal in the future. At the time, Galileo felt he was still on safe ground. He had powerful allies within the Church hierarchy, up to the Pope himself.

In a letter written December 1, 1611, Galileo confidently predicts the triumph "of the great Copernican system, to whose complete discovery such favorable winds assist, and such shining escort shows the way, that we need fear darkness and adverse storms no longer." Galileo had established himself as the lightning rod of the house of Copernicus. And swirling over the horizon was the "adverse storm" he had believed would never come.

Ironically, religious objections to Galileo's ideas first hatched in the academic world, most notably in that Aristotelian stronghold, Galileo's own alma mater, the University of Pisa. Leaflets and letters appeared that condemned Galileo. Galileo responded characteristically with his own barrage of acerbic letters and speeches. In 1616, Pope Paul V decided to settle the conflict. Despite a plea from Galileo that the Church keep separate the disparate realms of science and Scripture, the Pope declared the Copernican system contrary to the Holy Word. Copernicus's *De Revolutionibus* was placed on the Index of Forbidden Books, pending "correction." Galileo was enjoined from teaching that the Earth moves around the Sun, although he was free to pursue his research and discuss his Copernican beliefs in private forums.

Galileo laid the blame for the papal restrictions not on the Church, but on the conservative Aristotelian philosophers who had precipitated the Pope's action:

*They have endeavored to spread the opinion that such
[Copernican] propositions in general are contrary to the Bible
and are consequently damnable and heretical. . . . Contrary to
the sense of the Bible and the intention of the Church fathers,
if I am not mistaken, they would extend such authorities until
even in purely physical matters, where faith is not involved,
they would have us altogether abandon reason and the
evidence of our senses in favor of some biblical passage,
though beneath the surface meaning of its words this
passage may contain a different sense.*

And so it goes, even today, as religious factions try to wrest control of school science curricula, to wit, promoting biblical alternatives to the evolution of the species and the origin and development of the universe.

Galileo adhered to the papal prohibition, pursuing nonastronomical endeavors for the next decade. In 1623, Cardinal Maffeo Barberini, a long-time friend of Galileo's and a fellow member of the scientific Academy of the Lynxes, became Pope Urban VIII. With a kindred spirit now at the helm of the Church, Galileo decided it was safe to array the forces of Ptolemy and Copernicus against one another on the printed page. For the next six years, with thought and pen, he directed this clash of the world systems on his own private battlefield of ideas. In January 1630, Galileo completed *Dialogo . . . sopra i due massimi sistemi del mondo,* "Dialogue on the Two Chief World Systems." Italian historian Giorgio de Santillana describes *Dialogo* as "the story of the mind of Signor Galileo. But it is the mind of a man who knew very well where he was going. In the work there is all of him: the physicist, the astronomer, the man of the world, the litterateur, the polemicist, even at times the sophist; there is, above all, the totally expressive and expressed Renaissance man."

Galileo had come to realize that in the absence of stellar parallax, he could not *prove* the fundamental Copernican tenet that the Earth orbits the Sun. Instead, he decided to persuade his readers through logical argument, a technique he had honed his entire life. The linchpin of his argument was unrelated to his telescopic observations. Galileo proposed that the ocean tides could not arise if the Earth were stationary. He came up with an elaborate scheme whereby the Earth's motion, both in rotation and revolution, generated the movement of the waters. (He turned out to be wrong; decades later, Isaac Newton demonstrated that the Moon's gravity raised the tides.)

The Pope had given Galileo permission and even encouragement to write the book. But what Galileo delivered was nothing like Copernicus's *De Revolutionibus,* whose few readers—specialists all—had to gird themselves in mathematical armor, else be crushed by the book's complexity. No, *Dialogo* was entertaining, largely nontechnical, persuasive—and demonstrably pro-Copernican.

Dialogo portrays the lengthy and animated discourse among three individuals: Salviati, Galileo's alter ego; Sagredo, an open-minded Venetian nobleman, based on Galileo's friend of the same name; and Simplicio, the dimwitted Aristotelian goose who repeatedly stumbles over his own illogic. That the book's Copernican lessons pour from the mouth of a fictional character did little to disguise the teacher's true voice. It was Galileo's own. What's more, the text was in Italian. Any literate person could read it. Had Galileo set down the work in the usual Latin, the clerics might have been able to sweep the heretical ideas under the institutional rug of academia. But with *Dialogo* disseminated among the populace, conservative elements within the Church pressed to take action against its author. Soon they had convinced the Pope that behind *Dialogo*'s rhetorical façade was Copernican propaganda. Galileo might have complied with the letter of the 1616 papal prohibition, but certainly not with its spirit.

Church officials cited *Dialogo* as a violation of the prohibition against teaching the Copernican theory, this despite the fact that Church censors—and even the Pope himself—had initially approved its publication. (Galileo had consented to the censors' revisions before publication.) On April 12, 1633, Galileo, now nearly seventy years old and ailing, was brought before the Inquisition. Historian Santillana writes: "[H]e had gambled everything, not on deceiving but rather on persuading the leaders of the Church, or at least on giving them pause, before they hardened into their fatal decision [against the Copernican system].... He was a man of the Renaissance and a Christian of the old persuasion, to whom all this newfangled apparatus of thought police and propaganda brought about by the Counter-Reformation made little sense."

Under threat of torture, Galileo recanted his Copernican views. He was forbidden to publish any works in the future and was restricted to his house outside Florence for the rest of his life. Here in his enforced isolation, although suffering the effects of age and of the death of his beloved daughter Virginia, Galileo made sporadic astronomical observations and wrote his master work, *Discorsi e dimostrazioni matematiche intorno à due nuove scienze,* "Discourses on Two New Sciences." In *Discorsi,* Galileo details his conception of the properties of matter and the nature of motion. The manuscript, which anticipates Isaac Newton's mathematical explication of mechanics, was spirited out of Italy and published in Protestant Holland. By the time *Discorsi* appeared in 1638, Galileo had become totally blind. "This universe," he wrote to a friend, "which I with my astonishing observations and clear demonstrations had enlarged a hundred, nay, a thousandfold beyond the limits commonly seen by wise men of all centuries past, is now for me so diminished and reduced, it has shrunk to the meager confines of my body."

Had the parallax of even one star been detected in his time, Galileo's troubles might have been averted. Stellar parallax was the crucial piece still missing from the Copernican puzzle. Even without that piece, the overall Copernican picture was clear to thinkers like Galileo and Kepler, who filled in the breach by postulating the extreme remoteness of stars. But to others less forgiving, the absence of stellar parallax obliterated the Copernican picture entirely. To each "Simplicio" among this number, Galileo directs the following passage in his *Dialogo,* voiced by his mouthpiece Salviati:

> *I, with you, would say that, in case such a [parallax] were discovered, nothing more would remain behind that might render the mobility of the Earth questionable. But even if it should not sensibly appear, yet is not its mobility removed, or its immobility necessarily proved, it being possible (as Copernicus affirms) that the immense distance of the starry sphere renders such very small phenomena unobservable; these phenomena, as I have said, may possibly not have been hitherto so much as sought for, or, if sought for, yet not sought for in such a way as they ought, to wit, with that exactness which would be necessary to such a minute variation; and this exactness is very difficult to obtain, as well by reason of deficiency of astronomical instruments, subject to many alterations, as also through the fault of those who handle them with less diligence than is requisite.*

Not even the premier observer Tycho had succeeded in measuring the parallax of a star. But he had been working with his naked eye, as had every astronomer before him. No matter how well his instruments had been built or how carefully he had aligned them and read their graduated scales, Tycho had been hobbled by the inherent limitations of his own eye. The solution, according to Galileo's Salviati, was to "make use of instruments greater by far, and by far more certain than those of Tycho . . . an exquisite Telescope . . ."

In *Dialogo*, Galileo lays out both the groundwork and the methods by which future generations of astronomers, with telescopes more powerful and precise than his own, might measure the parallax of a star. The fundamental issue from the parallax hunter's standpoint is how to detect a minute shift in a star's position over a six-month interval—not *theoretically* how this process is to be executed, but the specific steps that must be taken at the telescope to give a reasonable chance of success. Galileo understood from the start that the procedure would not be as simple as affixing a "position scale" to the telescope, pointing the instrument to a star, and checking whether the position-reading changes over six months. No matter how finely divided such a scale, the parallax shift of a star is too small to be discerned in this way.

Galileo, through his alter ego Salviati, envisions two ways whereby stellar parallax might be measured. Galileo's first method is to fix a telescope securely to a frame or a post, such that it can point in only one direction. For reasons to be discussed later, a vertical orientation is most advantageous. Although the telescope is immovable, it is not stationary; the Earth's daily rotation sweeps the telescope's aperture across the starry sky. To the astronomer at the eyepiece, stars drift lazily into and across the circular field of view, as though the sky were a great conveyor belt conducting celestial objects into sight. Each star traces a line across the telescope's field. Some stars will happen to sweep through the middle of the field, others nearer the edge. Suppose the astronomer observes a star that sweeps precisely along the field's midline, such that the star passes through the exact center. If the Earth truly circles the Sun, then half a year later, our planet—and hence, the telescope it bears—will be far from the vantage point it had before. Remember, the telescope hasn't been adjusted in any way; it still points to where the star used to appear six months previously. Thus, the star's position will appear to have shifted slightly; having formerly tracked a midline course through the field of view, the star now passes slightly to one side. (To gain the maximum effect, the chosen star should be situated nearly perpendicular to the plane of the Earth's orbit.) Consider this Earthly analogy: If one photographs a subject across the room, then slides the camera laterally while keeping it pointed in the original direction and photographs again, the subject will appear shifted to one side in the second picture.

The shift between the star's current and former tracks through the telescope's field is attributed to the star's parallax, as viewed from opposite extremes of the Earth's orbit. If the star is not too remote, Galileo reasoned, and the telescope sufficiently powerful, the shift might be measurable.

Galileo's second method to detect stellar parallax begins by shattering the age-old myth of the celestial sphere. Salviati speaks of it to the receptive Sagredo: "I do not think that the stars are spread in a spherical surface equally remote from a common centre but hold that their distances from us are so various that some of them may be twice and thrice as remote as others." Here, Galileo has unceremoniously plucked the stars from the Aristotelian celestial sphere and strewn them throughout space. With this action, he has transformed the claustrophobic Ptolemaic universe of nested spheres into the modern Copernican universe of space. To what depth in the new cosmos do the scattered stars extend? The most distant, Galileo suggests in *Dialogo*, may be "twice and thrice" as remote as the nearest. But these numbers surely are meant as examples, not limits. The universe of stars might well extend infinitely, as far as Galileo is concerned. So for every "nearby" star, there might be countless others of unimaginable remoteness from the Earth. It is these faraway stars that provide Galileo the means to detect the parallaxes of nearby ones.

The parallax of a very remote star is negligible. As the Earth swings from one extreme to the other in its Sun-circling orbit, the remote star appears to stand stock-still, even when monitored in a powerful telescope. Thus, the star acts as a fixed marker in the sky, akin to the granite surveyor's stone that sits in one corner of my front yard and that demarcates my property from my neighbor's. The same is true of that star's equally remote cousins: every one, a luminous surveyor's stone, marking a precise point in the heavens. Taken together, these faraway reference stars form a fixed grid against which are measured the motions of closer celestial objects, such as planets, comets, or, in Galileo's case, nearby stars. But such reference stars are widely scattered in the sky. To measure the tiny parallax of a nearby star, Galileo realized that the reference star would have to be right alongside the nearby star in the telescope's field of view. As the Earth circles the Sun, the nearby star's parallactic wobble should stand out clearly in relation to its immobile "partner." In principle, Galileo's procedure is applicable to widely spaced stars; however, swinging the telescope from the "target" star in one part of the sky to a reference star in another part of the sky obliterates the measurement accuracy. No, for Galileo's method to work, target star and reference star must stand beside one another in the eyepiece.

There are many instances where two stars stand side by side in the sky. Astronomers call such pairs *double stars.* A notable example is the Mizar-Alcor pair in the "handle" of the Big Dipper. (Mizar is itself a double-double system, consisting of four stars altogether.) In a Galiliean-style telescope, Mizar and Alcor appear as two gleaming pinpoints, huddled together in the field of view. Some double stars are truly close to one another in space, held fast by their mutual tug of gravity. There are even cases where the gap between such physically associated stars is so small that their gaseous envelopes intermingle. On the other hand, many other double stars are fakes: Their proximity in the sky has no bearing on their proximity to each other in space. They are merely chance alignments, as seen from our unique Earthly perspective. In these false double systems, one star might lie relatively near the Earth, while its "partner" might lie very far away. This was precisely the type of double-star system Galileo claimed was amenable to his parallax method. But how could he tell which of the many double stars are pairings of a nearby and a faraway star?

Galileo was drawn to a particular subset of double stars: those that couple a very faint star with a very bright star. If brightness is any measure of a star's distance, he reasoned, then the faint member of the double must be far away, while the bright member must be relatively nearby. Galileo proposed that astronomers locate such bright-faint pairs and monitor the gap between the stars over many months. The parallax of the nearby, bright star should reveal itself in the cyclical widening and shrinking of the gap.

Galileo never tried to detect stellar parallax himself; he surely knew the difficulties of such an undertaking, not to mention the inevitable firestorm of criticism from the Church were he to succeed. He would leave it to others to transform his "exquisite Telescope" from a glorified spyglass into an engine of celestial measurement. Galileo was convinced that the elusive stellar parallax was real, yet so subtle that only a powerful, carefully constructed telescope could reveal it. He had fathered telescopic astronomy; he understood that the field was yet in its infancy and had to mature before this most difficult of observations stood any chance of succeeding.

It wouldn't be long before astronomers picked up Galileo's lead. One by one, they joined the effort to measure stellar parallax and claim the laurel of proving by direct observation the truth of the Copernican system. To this new generation of telescopic explorers, as to the naked-eye astronomers who preceded them, the finish line in the parallax race seemed just around the corner. Yet when they turned the corner, the finish was nowhere in sight. The sprint was becoming a marathon.

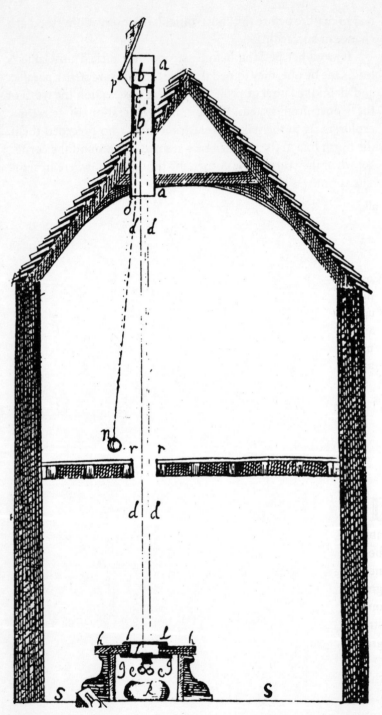

Robert Hooke's zenith telescope. From Hooke's 1674 paper on stellar parallax.
Source: *Owen Gingerich.*

8

The Archimedean Engine

If the study of the history of science is to teach us anything, we must make ourselves acquainted with the by-paths and blind alleys into which our forefathers strayed in their search for truth, as well as with the tracks by which they advanced science.
—*J. L. E. Dreyer, Tycho Brahe*

Coarse observations, made by honest and well-meaning men, have more perplexed the astronomer than all their labors and dreams upon them can make him satisfaction for. Their pretty thoughts and conceits in the theories are always excusable and sometimes to be commended: but when rude and ill-managed observations and experiments are brought to confirm them, though they may serve the author's present turn, yet they become a load on the science, and at last turn to his shame and reproach.
—*John Flamsteed, Astronomer Royal, 1685*

It is May 5, almost midnight. The sky is clear, the air comfortably warm through a sweater. Join me as I cut through my neighbor's yard, head down the gentle slope that serves in winter as the neighborhood sledding hill, and set up my telescope on the long grassy mall that everyone around here calls the "Aqueduct." The only telltale marks of the Aqueduct's human origins are its landing-strip straightness and the occasional incongruous manhole in the grass.

The Aqueduct is actually one of many canals of open land that cut across the cityscape outside Boston. The canal metaphor is apt because the Aqueduct is a tributary in a network of underground pipes that shuttle water from the Quabbin Reservoir in western Massachusetts to us thirsty Easterners, not unlike Percival

Lowell's fictitious Martian canals, which did the same for inhabitants of the Red Planet. Despite neighborhood sentiments to the contrary, the Aqueduct is controlled by the Massachusetts Water Resources Authority, which maintains it and posts the No Trespassing signs that are routinely ignored or taken down. Much of the adjacent suburban construction sprouted after the aqueduct system was laid and hugs these grassy runways so tightly that it looks as though a mammoth lawn mower cleared linear swaths across metropolitan Boston.

The Aqueduct is the only place near my home where a telescope has a clear shot at much of the night sky, although suburban glare obliterates many of the fainter celestial sights. Ironically, the more astronomers roll back the frontiers of the universe, the less of it the ordinary citizen is able to see because of local light pollution. Tonight's target, however, is sufficiently bright to penetrate this luminous interference: Gamma Draconis, also known as Eltamin, a moderately bright star in the "head" of Draco the dragon. The name Eltamin derives from *Al Ras al Tinnen*, "the Dragon's head," by fifteenth-century astronomer Ulugh Beg, grandson of the Mongol conqueror Tamerlane. The more prosaic designation, Gamma, is first encountered in the beautiful *Uranometria* sky atlas drawn by German celestial cartographer Johannes Bayer in 1603. In Bayer's system, a constellation's brightest star is labeled Alpha, the second brightest Beta, and so on. Like its mythical reptilian namesake, Gamma Draconis has had a long affiliation with human affairs. At least one ancient culture believed that any passage of a comet near this star was an ill omen. Gamma Draconis also played a significant role in Egyptian religious and mythical traditions. The central passageways of seven Egyptian temples, including the 1,500-foot-long corridor of the great Ramses monument at Karnak, are aligned toward the horizon point where the star rose each night around 3500 B.C. Because of the slow, toplike gyration, or precession, of the Earth's axis, Gamma's rising point on the horizon shifts gradually with time, a fact reflected in the slightly differing temple alignments over the centuries during which they were built. In fact, each temple can be dated "astronomically" by its individual alignment.

Tonight, in the eyepiece of my small telescope, Gamma Draconis appears as a fiery, yellow-orange point. Radiant spikes boil up angrily from the starry image, as though it were a cat trying to claw its way out of a luminous sack. It's easy to become mesmerized by the scintillating image of a star in the telescope. Its

light floods my eye, a silent siren's call that draws my consciousness deep into space. Soon I sense a connection between me and
the star, as though its beam were a searchlight that had turned my
way and was illuminating my mind.

Gamma Draconis looks pretty much like any of the thousand
or so other stars above my head tonight. Technically, Gamma
Draconis is a "giant" star. It's about 25 times wider, although
somewhat cooler, than the Sun, and radiates into space about 180
times more energy. It lies approximately 150 light-years from the
Earth—900 trillion miles. The light now streaming into my telescope began its journey about the time Marx and Engels penned
the last line of the Communist Manifesto and a young Mark
Twain dreamed of becoming a Mississippi River pilot. Tonight, as
on any night, only an infinitesimal fraction of the star's photons
are swept up by my telescope. Others sprinkle the Aqueduct
around me like fairy dust, their luminous energy dissolving into
the ground as minuscule pulses of heat. Still others leave Gamma
Draconis in directions that carry them nowhere near the Earth;
some of these might be streaming into a telescope or sprinkling
an aqueduct on another world. Nature is profligate with its starlight. Most is not destined for sentient consumption, but falls
dumbly onto dust grains in outer space or flies unhindered through
the void.

At this hour, from my vantage point on the Aqueduct,
Gamma Draconis lies in the eastern part of the sky. It is not yet
halfway up from the horizon and must still contend with some of
the taller trees around me. Roughly four hours from now, it will
have risen to its maximum altitude for the night: at Boston's latitude, that's about 80 degrees above the horizon, or 10 degrees shy
of straight over my head. Then the star will sink slowly toward the
west until it sets. The cycle will repeat the next night.

While the ancient Egyptians had held Gamma Draconis in
high esteem, the star's name stirs no recognition among the modern public. Yet Gamma Draconis is one of the most significant
stars in all of astronomical history. Its fame has nothing to do
with any intrinsic property. It is neither the brightest nor the
largest star, neither black hole nor supernova. It is not one of the
select stars around which planets have recently been observed. No
poet, famous or obscure, pens odes to this star. Although classified
as a "giant," Gamma Draconis is nonetheless rather ordinary, an
everyman of the stellar multitude. The source of the star's unlikely
notoriety lies in an accident of placement: Randomly situated as

it is in the heavens, Gamma Draconis passes almost precisely overhead in London once every day. This circumstance is far from unique; at every location on the Earth, some star passes exactly overhead during the course of the night. In Boston, one such example is the variable star Algol, Medusa's glittering eye, in the constellation Perseus. That Gamma Draconis passes overhead in London mattered not a whit from the dawn of civilization until well into the second half of the seventeenth century. Then scientist Robert Hooke proposed an adaptation of the experimental method suggested by Galileo: An overhead star might be studied with a fixed, vertical telescope to detect stellar parallax. Hooke also happened to be the one scientist in England most capable of constructing such an instrument, and he had recently taken up residence at London's Gresham College. Once every day, he noted, the star Gamma Draconis wheeled directly overhead. In 1669, Robert Hooke cut a hole in the roof of his apartment, a hole through which he could point a telescope—straight up.

Robert Hooke was a sickly child practically from the day he was born on the Isle of Wight in 1635. His parents didn't expect him to survive, but he did, along with the chronic headaches, stomach upsets, sleeplessness, and, as he later described them, "wild frightfull dreames" that dogged him throughout his life. Because of his fragile health, Hooke was schooled initially at home by his father, a minister. Upon his father's death, thirteen-year-old Hooke continued his education at the Westminster School before enrolling at Oxford in 1653. Although he never completed a bachelor's degree, Hooke impressed several energetic junior faculty, including Christopher Wren and Robert Boyle, who paved the way for Hooke's eventual career in science.

At least by seventeenth-century standards, Robert Hooke was an unusual-looking man. The ever-perceptive Samuel Pepys tidily sums him up in a diary entry dated February 15, 1664: Hooke "is the most, and promises the least, of any man in the world that I ever saw." The rumored sole portrait of Hooke, if it ever existed, was reportedly disposed of by a vengeful Isaac Newton after Hooke's death. Nevertheless, contemporary descriptions have painted a vivid picture of Hooke in my imagination as a cross between Ichabod Crane and the Hollywood version of Dr. Frankenstein's devoted assistant, Igor. Hooke's posture began to deteriorate when he was sixteen, perhaps from a scoliotic spine, so that within a few years he was unable to stand completely erect. Hooke's friend, the chronicler John Aubrey, describes him as "but of middling stature, something crooked, pale faced, and his face but little below, but his head is lardge; his eie full and popping, and not quick." Then he adds: "Now when I have sayd his Inventive faculty is so great, you cannot imagine his Memory to be excellent, for they are like two Bucketts, as one goes up, the other goes downe." Richard Waller, secretary of the British Royal Society and Hooke's first biographer, notes that Hooke kept his hair "very long and hanging neglected over his Face, uncut and lank."

A committed diarist himself, Hooke kept tally of the interplay between his diet and his fluctuating state of health, as in these characteristic entries: "19th December 1673 Apples agreed well. *Deo Gratias.*" "29th January 1676 Chocolate stone-sturgeon vinegar, agreed not." "7th August 1678 Poisoned by sour beer." He was a habitual chocolate drinker, a vice he tried to forswear—unsuccessfully. Among his possessions auctioned at his death, a book: *The Virtue of Chocolatt.* To relieve his ever-present symptoms, Hooke dosed himself regularly with medicinal herbs and minerals, including iron, mercury, lignum vitae, laudanum, rhubarb, and wormwood. He took sal ammoniac, in his words, to "dissolve that viscous Slime that hath soe tormented me in my stomack and gutts." Hooke biographer Richard S. Westfall notes that "Hooke's spiny character was nicely proportioned to the daily torment of his existence." So knowledgeable did Hooke become about bodily ills and medicinal cures that he was awarded an M.D. degree in 1691. Hooke never married, but carried on sequential affairs with a number of housemaids and ultimately with his niece, Grace, who had lived in his home from age eleven.

Hooke compensated for his physical limitations with a seemingly inexhaustible nervous energy. Rather than retreat into the background, as others in his circumstances might have done, Hooke thrust himself into the limelight, hunched back, hypochondria and all. "He went stooping and very fast," his contemporary, Richard Waller, tells us, "having but a light Body to carry, and a great deal of Spirits and Activity." Despite his unconventional appearance and frequently stormy nature, Hooke won the respect and friendship of many. His diaries describe a full social calendar, whether he's spurring spirited conversations at parties or downing a few pints at the pub with his compatriots. On the other hand, Hooke carried on interminable disputes with the astronomers Johannes Hevelius and John Flamsteed and with Isaac Newton, among others. (Hooke on Flamsteed: "an ignorant impudent ass." Flamsteed on Hooke: "much troubled with Mr. Hooke who . . . force[s] his ill-contrived devices on us.") Hooke was much impressed with himself and apparently found it hard to accept that others might be less so.

Robert Hooke's forte was experimentation and mechanical innovation. He displayed extraordinary mechanical abilities, even in childhood. As a youth, he constructed a fully rigged, yard-long man o'war with working cannon and also a functional clock entirely of wood. According to Aubrey, the young Hooke "invented thirty severall wayes of Flying," but abandoned the effort after concluding that muscle power was insufficient to realize this flight of imagination.

Hooke's scientific career took off in 1662 when his Oxford circle secured his appointment as Curator of Experiments of England's recently chartered Royal Society. His job was daunting by any measure: "to furnish the Society every day they meete [which was typically once a week], with three or four considerable Experiments, expecting no recompense till the Society gett a stock enabling them to give it." For the next fifteen years, Hooke generated an uninterrupted stream of inventions and experiments for Royal Society members to peruse and discuss. The weekly gatherings frequently became a forum on Robert Hooke's creations, without which the Society might have withered. The meeting descriptions for the 1660s in Thomas Birch's *History of the Royal Society* are peppered with phrases like "Mr. Hooke proposed," "Mr. Hooke produced," "Mr. Hooke was ordered," "Mr. Hooke was put in mind," "Mr. Hooke remarked," "Mr. Hooke related," and "Mr. Hooke made some experiments." Even in the wake of London's Great Fire of 1666, when Hooke added to his workload the architectural and inspectional duties of City Surveyor, the "considerable Experiments" continued to flow. Such was Hooke's lifestyle that his chronic insomnia proved advantageous.

Like a latter-day Leonardo, Hooke established himself as one of the seventeenth century's most creative and prolific inventors. But while Hooke's illustrious predecessor rendered on paper a multitude of fantastic machines, Hooke actually built many of his. Among the thousand or so new devices and processes for which Hooke took credit are an improved air compressor and vacuum pump; the spiral balance spring, which allows timepieces to be used in any orientation; the fundamental instruments of meteorology, including the wheel barometer, hygrometer, and wind gauge; the universal joint; a mechanical calculator for multiplication and division; the iris diaphragm, eventually incorporated into cameras; mercury amalgam, later used in dental fillings; a clock-driven telescope that automatically keeps celestial objects in view while the Earth turns; and the mathematical equation of spring action, known to every freshman physics student as Hooke's law. (Upon learning of Hooke's expertise in this last area, Charles II asked for a large spring scale he could use to weigh himself after tennis.)

In 1664, Hooke became Professor of Geometry at Gresham College, where the Royal Society held its meetings. The college had been founded in the late 1500s by wealthy merchant Thomas Gresham as a wellspring of enlightenment for the London citizenry. In the founder's mansion on Bishopsgate Street lived seven faculty members who performed scholarly work and delivered public lectures in the arts and sciences. Hooke moved into a two-story corner apartment overlooking Gresham College's enclosed quadrangle. About the same time, another wealthy merchant, Sir John Cutler, tapped Hooke for one of his eponymous lectureships. Cutler was notorious for bestowing such honorifics without following through with the cash. But Hooke did receive his paycheck—thirty-two years later, after bringing suit against Cutler.

Robert Hooke vaulted to international prominence with the publication in 1665 of *Micrographia,* a richly illustrated tour of the world of the diminutive. What Galileo's *Sidereus Nuncius* did for astronomy, Hooke's *Micrographia* did for microscopy. The public was riveted by Hooke's sober, yet unearthly, renderings of a fly's eye, a snail's teeth, mold stalks on a leaf, a razor's edge, frost figures in water and urine, eels in vinegar, plus a host of pinhead-sized arthropods—gnat, flea, mite, and louse—all magnified to nightmare-inducing proportions. So brimming was Hooke with new ideas that he appended to his book a potpourri of observations and musings that have little or nothing to do with the microscope. There are engravings of a mercury barometer, a lens-grinding machine, the Moon's surface as seen through a powerful telescope, and the Pleiades star cluster. Hooke also describes a precursor to the wave theory of light, experiments to prove that lunar craters are of impact origin, and hypotheses about combustion and the structure of crystals. Samuel Pepys pronounced *Micrographia* "the most ingenious book that I ever read in my life." Isaac Newton later praised it during a Royal Society presentation.

Hooke was a one-man idea factory, whose ceaseless output provided the opportunity for enlightenment. Or for trouble. He was as possessive of his own inspirations as a bulldog of a bone; even if one manages to steal the bone, there are the jaws to contend with. And with minimal provocation, Hooke's tendency was to snap. Lacking the time or inclination to pursue every one of his insights, Hooke left many by the wayside. Inevitably a colleague would adopt one of the orphaned ideas, or a competitor would conceive a similar idea independently. Just as inevitably, Hooke would accuse these scientists of capitalizing on his efforts.

Hooke's wariness was not unfounded, given the tenor of the times. Research journals were only being introduced, and the dissemination of scientific information was carried out primarily by letter or by word of mouth. Establishing one's priority for an invention or an experimental finding was difficult. Many scientists, Galileo and Hooke among them, encoded their discoveries in complex anagrams that they mailed to colleagues. A mistranslation of one such anagram had led Kepler to believe that Galileo has discovered two moons circling Mars, when in fact Galileo's message described observations of Saturn's "companions," or as we know today, its rings. (By coincidence, Mars was later found to have two moons, neither of which would have been visible in Galileo's telescope.)

In Hooke's time, John Aubrey castigated mathematician John Wallis: "[He] steals flowers from others to adorne his owne cap,—e.g. he lies at watch, at Sir Christopher Wren's discourse, Mr. Robert Hooke's, Dr. William Holder, &c; putts down their notions in his note booke, and then prints it, without owneing the authors. This frequently, of which they complain." Hooke commented in his own diary: "Shewd my quadrant to all but Oldenburg"; "told Sir Robert Southwell that I could fly, not how."

The most famous example of Hooke's professional jealousy gone awry, and the one that probably condemned him to relative obscurity, was his long-running dispute with Isaac Newton. As accomplished as he was, Robert Hooke never approached the analytic genius of his younger competitor from Cambridge. Nor did Hooke possess Newton's legendary tenacity when it came to solving a thorny problem. (Newton's reply, when asked how he had derived his laws of gravity and motion, "By thinking of them without ceasing"—an answer that can be taken quite literally.) Unlike Newton, Hooke's rapid-fire imagination propelled him from one project to the next, like a sailboat driven helplessly before a shifting gale of ideas.

Hooke had stoked the fires of Newton's enmity. He openly hurled barbs at Newton's work. For instance, after Newton displayed his revolutionary reflector telescope to the Royal Society in 1672, Hooke sniffed that the little lens he carried around in his pocket was better. Hooke accused Newton of appropriating his ideas about the nature of light and of gravity. Specifically, it was Robert Hooke who first envisioned both that gravity was a mutual interaction between bodies and that the nature of gravity on the Earth is identical to that between celestial bodies. It was Hooke, among others, who proposed that the gravitational tug between two bodies that are moving apart diminishes with the square of their separation, the so-called inverse square law. Hooke also anticipated Newton's principle of inertia, whereby a planet, which in the absence of gravity would proceed through space in a straight line, is deflected into orbit by the Sun's gravitational pull. Newton undoubtedly received inspiration from Hooke on key scientific points and, had the two men not had such a stormy history, might have acknowledged that debt. But as it was, Newton expunged every reference to his antagonist in the manuscript of his great work, the *Principia*. It must have been doubly mortifying to Hooke that, with his limited training in mathematics, he himself was incapable of proving his own speculations. All he could do was stand by while Isaac Newton created the mathematical masterpiece that formed the bedrock of classical physics.

The ongoing animosity between the two men percolated for decades, a grim situation that prompted poet John Dryden to remark, "Two such as each seem'd worthiest when alone." In the dispute over who had first suggested the inverse square law of gravitational force, John Aubrey rallied to his friend Hooke's defense, ascribing to him "the Greatest Discovery in Nature that ever was since the World's Creation. It never was so much as hinted by any man before." Nevertheless, Aubrey chides Hooke for not pre-empting this particular dispute with Newton: "I wish he had writt plainer, and afforded a little more paper."

Hooke ultimately made seminal contributions in areas as diverse as astronomy, physics, meteorology, microscopy, horology, and geology. It is no exaggeration to say that Hooke's universe of expertise ranged from the hairs on a flea's back all the way out to the stars. It is only in Newton's ponderous shadow that Hooke's reputation lies diminished. If it is hard today to associate Robert

Hooke with any single discovery, it is only because he explored so many different frontiers—or as he himself confidently noted in his diary on June 28, 1680: "Spent most of my time in considering all matters."

Before the assembled Royal Society in 1669, Robert Hooke, Curator of Experiments, announced his intention to begin a most ambitious foray into astronomy. In fact, it would be the most difficult celestial observation ever undertaken. But if successful, he would join the pantheon of astronomical luminaries like Ptolemy, Copernicus, Tycho, and Galileo. Hooke was planning to measure the first parallax of a star. And he already knew which star it would be.

"Whether the Earth move or stand still," Hooke wrote, "hath been a Problem, that since Copernicus revived it, hath much exercised the Wits of our best modern Astronomers and Philosophers, amongst which notwithstanding there hath not been any one who hath found out a certain manifestation either of the one or the other Doctrine." With naïve determination, Hooke laid out his proposal to prove the Copernican theory by direct observation. At first, the Royal Society members were dubious, finding Hooke's plan "somewhat extravagant, and hardly practicable." But as they listened further to their talented Curator of Experiments, the odds of success no longer seemed quite as remote.

Hooke planned to use a fixed, vertical telescope of high magnification. A vertical telescope, he explained, has two advantages. First, gravity defines the vertical direction unambiguously; by aligning the telescope parallel to a plumb bob, the instrument will always point straight up to the "top" of the sky. This zenith point would be the reference marker from which to measure the position of a given star—and to detect any variation in that position during the course of the year. The second advantage of a vertical telescope is the absence of atmospheric refraction. Every light beam that threads the atmosphere from outer space bends, or *refracts*, as it traverses the ocean of air, just as a ray of light bends when passing through a block of glass, a spectacle lens, or a beaker of water. In the case of cosmic light, the more obliquely the beam strikes the atmosphere, the more it is refracted. By the time the beam from a star arrives at the observer's eye, its direction has been altered by as much as half a degree, or a full-Moon's width. Thus, to the eye, the star appears slightly higher in the sky than it truly is; that is, higher than its perceived position were the atmosphere absent. When we see the Moon "resting" on the horizon, it is actually *below* the horizon; light from the obstructed Moon has been refracted such that the lunar image becomes visible. The same is true for the rising or setting Sun.

Atmospheric refraction would wreak havoc in a long-term effort to detect the tiny position shifts caused by parallax, especially if the star was observed at different places in the sky from night to night. Fortunately, there is one situation in which the refractive effect of the Earth's atmosphere is effectively nullified: when the starlight penetrates the atmosphere straight down, not at an angle. From the observer's point of view, such a star lies directly overhead at the zenith. For light entering the atmosphere straight down, Hooke realized, "be the Air thicker or thinner, heavier or lighter, hotter or colder, be it in Summer or Winter, in the night or the day, the ray continually passeth directly, and is not at all refracted and deflected from its streight passage."

Hooke told the Royal Society that a bright star did indeed exist near the zenith of Gresham College: Gamma Draconis. Although Gamma Draconis did not pass *exactly* through the zenith, where Hooke's vertical telescope would be pointing, it nevertheless passed close enough to fall within the instrument's narrow field of view. Hooke's plan was straightforward: Wait for Gamma Draconis to arc overhead and into the sight of his telescope, then measure how close the star passed to the zenith on that given night. If the Earth truly circles the Sun, as Copernicus had held, that "zenith distance" would change cyclically over the year due to the star's parallax.

As was his nature, Hooke derided those who adhered to the Ptolemaic or Tychonic world systems or who refused to accept the gargantuan proportions of Copernicus's universe. He characterized non-Copernicans as those "conversant only with illiterate persons," and who considered "the Sun as big as a Sieve, and the Moon as a Cheddar Cheese, and hardly a mile off." He was equally critical of astronomers who refused to acknowledge the superiority of telescopic sights when measuring star positions. In this arena, he stood head-to-head with the noted observer Johannes Hevelius in Danzig. Hevelius did use telescopes to study the heavens, but when measuring star positions, he preferred the old-fashioned, Tycho-style quadrants and sextants, where the observer measures each star position by eye. Hooke proved experimentally that naked-eye sighting had reached its practical limit of precision—about one arcminute, or $1/60$ of a degree—and that more finely determined star positions could only come from telescope-assisted measurement. Hevelius countered by claiming that telescopic sights were nearly impossible to align properly, and that as a result the observer was always reading incorrect star positions. (Like a misaligned rifle sight, even if the target is centered in the sight, the bullet will never strike the bull's-eye.) The situation became so heated that the Royal Society dispatched astronomer Edmond Halley to mediate the dispute, but Hooke and Hevelius ultimately went to their graves believing the other was wrong.

Hooke named his revolutionary telescope the Archimedean Engine, an allusion to the ancient philosopher's claim that, given a long-enough lever and a place to stand, the Earth might be moved. Now, in 1669, it was Robert Hooke's turn to pry loose our planet from its Ptolemaic shackle and send it spinning in its proper Copernican orbit.

Hooke began by cutting a passage, about a foot square, through the roof and ceiling of his Gresham College apartment. Into this hole he inserted a vertical, ten-foot-long square tube, such that the lower end poked into the upper-story room below. He cut a second hole in the floor of this room, over which he laid a set of shutters. With the shutters open, celestial light had an unobstructed, thirty-six-foot-long vertical path from the entry point on the roof to the observing station on the apartment's first floor. Into the roof tube, Hooke secured a lens of thirty-six-foot focal length. To protect the lens from the elements when not in use, he covered the tube with a hinged lid that could be actuated by pulling a string within the apartment. Hooke would observe Gamma Draconis through an eyepiece mounted in a frame on the first floor. Underneath the eyepiece was a couch on which Hooke would lie while observing.

Suspended from the bottom of the roof tube was a pair of long, silk threads, each bearing a lead ball, or "plumbet." Each ball rested in its own small water trough on the first floor to dampen the effects of vibrations and air currents. The weighted silk threads indicated the vertical direction on a graduated scale adjacent to the eyepiece. All Hooke had to do was await Gamma Draconis's arrival in the eyepiece each evening, point the telescope directly at the star, then record the deviation of the telescope from the vertical plumb lines. The eyepiece holder could be slid sideways, such that Hooke could observe stars up to half a degree from the zenith. In this way, he'd have advance warning of Gamma's approach.

Hooke made his first observation of Gamma Draconis on July 6, 1669, finding that the star passed north of the zenith by just over two arcminutes, about $1/30$ of a degree. He repeated the observation on July 9 and August 6, with nearly identical results. This was no surprise to Hooke, given that over a month's time, the Earth might not have moved enough to make the star's parallax noticeable. For his next observation, Hooke waited until October 21. On this date, Gamma Draconis was due to pass overhead, not during the night, but in the middle of the afternoon. Even in broad daylight, Hooke was able to see the star clearly in his telescope. (This is true for any bright star or planet, as long as you know precisely where to point your telescope.) At 3:17 P.M., Hooke watched Gamma Draconis pass somewhat less than two arcminutes north of the zenith, slightly different from the three prior observations.

Here, possibly, was the first detection of stellar parallax. Follow-up observations to confirm the effect were now critically important. Yet Hooke summarily abandoned the project: "Inconvenient weather and great indisposition in my health, hindred me from proceeding any further with the observation that time, which hath been no small trouble to me." No small trouble, indeed. In his treatise *An Attempt to Prove the Motion of the Earth from Observations,* published in 1674, Hooke describes the many tribulations that beset him trying to ensure that his telescope worked reliably. The instrument invariably fell out of alignment and had to be readjusted. The roof tube warped in various directions depending on wind and weather, and shifted the orientation of the telescope's main lens. Hooke's frustration led him to wonder whether a similar telescope might be mounted instead within a deep, dry well, such as the one he had seen "at a Gentlemans house not far from *Banfield Downs* in *Surry,* which is dug through a body of chalk, and is near three hundred and sixty foot deep." An abortive attempt to observe Gamma Draconis using a well telescope was in fact carried out in 1679 by astronomer John Flamsteed on the grounds of the Royal Observatory at Greenwich. A period etching by Francis Place indicates that the well was about a hundred feet deep and was fitted with a thirteen-turn spiral staircase and a "pineapple" cupola. The well no longer exists, although the telescope's ten-inch-wide objective lens is on display in the Science Museum in London. Flamsteed can be forgiven for abandoning the project "because of the damp of the place." Hooke and Christopher Wren also toyed with the idea of mounting a zenith telescope within London's Monument, the 202-foot-tall Doric column that Hooke designed (although it is typically credited to Wren) after the Great Fire. But upon the monument's completion in 1676, Hooke determined that vibrations within the hollow column would have rendered such a telescope ineffective.

By October 21, 1669, when he made his final observation of Gamma Draconis, Hooke must have wearied of the project. Its theoretical foundation was sound, but its implementation had become an escalating nightmare. Hooke certainly realized that a successful parallax measurement would probably involve radical improvement of his apparatus, if not a new apparatus altogether. In any event, the Archimedean Engine came to an ignominious end: Its lens broke. "An unhappy accident," Hooke described the misfortune without elaboration.

Having lost the stellar parallax battle, Robert Hooke nonetheless decided to declare victory. Colleagues tried in vain to convince him that his measurements were too few to be conclusive. Prominent in his 1674 treatise is the announcement: "'Tis manifest then by the observations . . . that there is a sensible parallax of the EarthsOrb to the fixt Star in the head of *Draco,* and consequently a confirmation of the Copernican System against the *Ptolomaick* and *Tichonick.*" Hooke estimated the parallax shift of Gamma Draconis to be roughly thirty arcseconds, or somewhat less than $1/100$ of a degree.

As critics contended and as time proved, the number was entirely spurious. Like many astronomers who preceded him and others who would follow, Hooke's preconception about the size of stellar parallaxes was hopelessly inflated. He had duped himself into believing that he had observed a star's parallax, when in fact he had seen nothing except the shadows cast by his own expectations. By modern measurement, Gamma Draconis's parallax is less than $1/1,000$ of what Hooke had held it to be. The Archimedean Engine, while clever in design, was much too blunt an instrument to detect the subtle shifting of the stars. And Robert Hooke, while also clever, lacked the determination to elevate the investigation to the next level. On the technological side, more refinement was needed; on the human side, more patience.

Patience was the watchword for the Reverend James Bradley, born in Sherbourn, England, in 1693, and like his restless predecessor Robert Hooke, an Oxford man. From his vicarage at Bridstow in Monmouthshire, James Bradley quietly tended his flock of parishioners during the day. At night, however, while his congregation slept, pulpit gave way to telescope, vestments to warm coat, and prayer to observation. When James Bradley raised his eyes to the heavens after dark, it was frequently not as a servant of God, but as an agent of science. In 1721, just two years into his vicarage, Bradley surrendered himself to his life's calling. And it was not religion.

James Bradley.
Source: *Wolbach Library, Harvard University.*

9

A Coal Cellar with a View

*Unquietness, anxiety, dissatisfaction and torment, these are
what nourish science. Without fundamental anxiety,
there is no fundamental research.*
—Jacques Monod, *Nobel Prize winner in medicine, 1965*

Vision is the art of seeing things invisible.
—Jonathan Swift, Thoughts on Various Subjects, *1711*

★ ★ ★

When Handel's *Water Music* began to play on the radio this
morning, the now-familiar image of the Reverend James Bradley
immediately popped into mind. For some reason, the sight of this
jowly, bewigged man with the Mona Lisa gaze stirs hunting horns
and trumpets inside my head. Perhaps it's because in the late
1720s, when Bradley carried out his extraordinary observations,
Europe was ringing with music whose robustness and technical
virtuosity were unlike anything heard before: Georg Friedrich
Handel composing operas in London, Johann Sebastian Bach
penning the score of his *St. Matthew Passion*, Antonio Vivaldi
standing at the apex of his career, Bartolomeo Cristofori perfect-
ing the mechanical action of his innovative keyboard instrument,
the pianoforte. It was a new age in astronomy as well. The novel
but technically primitive telescopes of Robert Hooke and his contemporaries were giving way to ever more sophisticated instruments: true engines of celestial measurement, sturdily mounted,
exquisitely balanced. And in the 1720s, no one was better suited to
take advantage of such instruments than the Reverend James
Bradley.

In the hunt for stellar parallax, James Bradley would follow
directly in Robert Hooke's footsteps. Yet the two men could not

have been more different. If Hooke was the eruptive nova, Bradley was the steadfast, quiescent Sun. Hooke had made the entire realm of science his hunting ground; Bradley devoted his energies almost exclusively to astronomy. Hooke had flitted from inquiry to inquiry, as his momentary passions suited him; Bradley, on the other hand, pursued tedious, multiyear investigations with Buddha-like aplomb. When it came to research, Bradley displayed an almost preternatural tenacity that prompted historian Agnes Clerke to remark, "The slightest inconsistency between what appeared and what was to be expected roused his keenest attention; and he never relaxed his mental grip of a subject until it had yielded to his persistent inquisition."

The contrast between Robert Hooke and James Bradley extends beyond their professional styles. Hooke's personal life had been tumultuous and rife with heartache; Bradley's was singular in its placidity. "I think we shall be inclined to admit," wrote nineteenth-century astronomer Herbert Hall Turner, "that if ever there was a happy life, albeit one of unremitting toil, it was that of James Bradley." Hooke had engaged in multiple romantic liaisons; Bradley appears to have had no romantic attachments before age fifty-one, when he married. Hooke had fostered prolonged feuds with colleagues; Bradley seems to have stirred no enmity among his peers, being "peculiarly kind and gentle in his manner." Indeed, the Earl of Macclesfield, in nominating Bradley to the post of Astronomer Royal in 1742, wrote that Bradley's "character in every respect is so well established, and so unblemished, that I may defy the worst of his enemies (if so good and worthy a man have any) to make even the lowest or most trifling objection to it." Or contrast Isaac Newton's well-known disdain for Robert Hooke with his pronouncement that James Bradley was "the best astronomer in Europe."

I can easily picture Robert Hooke's stooped, energetic figure bustling through the streets of London, as though perpetually five minutes late. On the other hand, contemporary reports paint Bradley as a composed, "chin-up" presence, who would have strode at a deliberate pace—not regally, mind you, for he was by all accounts a modest man. As an educator, I feel a certain kinship with Bradley the teacher. While never praised as an inspirational lecturer, Bradley was said to be responsive to his students, patiently explaining abstruse concepts. As to the other educational role that we share—interacting with the general public—here too I sense a connection with my eighteenth-century predecessor. From years of experience, I can reliably predict the course of conversation after being introduced to someone at a social gathering. Once the preliminaries are dispensed with, talk invariably turns to one's profession: *And what do you do for a living? Oh, you're an astronomer. I've never met an astronomer before.* Then come the questions about black holes or flying saucers or why the Moon looks so big when it's on the horizon. I suspect that conversation was little different in Bradley's time, when the general public must have wondered about the arrangement of the heavens and what stars are made of and why the Moon looks so big when it's on the horizon. The reticent Bradley was "averse from the promiscuous conversation of common society," but once engaged, would graciously "adapt himself to the capacity and understanding of those he conversed with."

Bradley learned astronomy from his uncle, Reverend James Pound, rector at Wanstead, near London, with whom Bradley frequently stayed. Born in 1669, the same year that Robert Hooke pointed his Archimedean Engine toward the heavens, Pound had preceded his nephew James to Oxford, although "his record indeed was such as good tutors of colleges frown upon." For all of thirteen years, Pound had flitted among different areas of study, including medicine, before seeking adventure in 1700 as chaplain to the settlement on the island of Poulo Condore, off the coast of Vietnam. He returned to England in 1706, one of a handful of survivors of a native insurrection. Where or how James Pound picked up the art of astronomical observation is a mystery. Young Bradley adored his Uncle James, who helped support him financially, nursed him through smallpox in 1717, and ultimately fostered his love of astronomy.

By the time Bradley was in his twenties, he and his Uncle James had formed a for-hire observing partnership. So respected were their skills that both Isaac Newton and Edmond Halley, soon-to-be Astronomer Royal, entrusted them on multiple occasions with observing projects. Bradley and Pound determined with unprecedented accuracy the positions of stars and nebulae, observed eclipses of Jupiter's satellites, and measured the diameter of Venus (with a 212-foot-long telescope) and also the parallax of Mars. Bradley himself calculated the orbits of two comets from repeated observations of their positions in the sky.

In 1721, James Bradley informed his congregation at Bridstow that he was leaving. With a recommendation from Isaac Newton, he had been appointed Savilian Professor of Astronomy at Oxford. Oxford had long had astronomers on its faculty, although it had little in the way of observing equipment. Given his modest annual salary of £140, Bradley could not afford to live at the university. Instead he moved in with his Uncle James in Wanstead and visited the Oxford campus only to deliver the required lectures.

In 1724, two events occurred that altered the course of Bradley's career: the first was the death of his beloved uncle; the second, the entry of a new observing companion, Samuel Molyneux, a wealthy amateur astronomer and member of Parliament from Kew, outside London. Having read of Robert Hooke's failed attempt to detect stellar parallax, Molyneux had set himself the lofty goal of repeating the experiment at his private observatory, but with better equipment. All he needed now was the guidance of an expert. Thus began James Bradley's observations of England's "overhead" star, Gamma Draconis, observations that led to one of the most significant and unexpected discoveries in the history of astronomy.

In 1725, Molyneux commissioned from Fleet Street artisan George Graham a custom, high-power zenith telescope, similar in principle to Hooke's Archimedean Engine. Graham was England's foremost telescope and precision clock maker, who according to a contemporary report was "the best mechanician of his time." The zenith telescope Graham fabricated had a tin plate tube, twenty-four feet long and slightly under four inches in aperture. Molyneux attached the instrument vertically to the face of a chimney in his mansion bordering Kew Green. To accommodate the telescope, holes were cut through the roof and between floors. The upper end of the tube was secured to an iron pivot such that the lower end, containing the eyepiece, could be swiveled slightly north or south by turning a micrometer screw. The resultant tilt of the telescope was read off a graduated arc, whose zero-point—corresponding to the vertical—was indicated by a plumb line hanging from the top of the telescope. If the tube was parallel to the plumb line, the telescope was pointed directly at the zenith.

For each nearly overhead passage of Gamma Draconis, the telescope's tilt was adjusted such that the star's natural east-west progression took it through the center of the view field. At that instant, when the star was at its maximum altitude for the night, the tilt reading was recorded. Graham had even inserted fine wire crosshairs into the eyepiece to help in centering the star. Every element of the telescope's design was optimized to a single end: to measure the parallax of Gamma Draconis. (English astronomer William Gascoigne got the idea for the crosshair around 1640 after a spider had spun a web in the tube of his telescope; the web stood in clear focus against the stars in the view field. Tragically, the inventive Gascoigne was killed in 1644 during the English Civil War. He was twenty-four years old.)

Bradley knew what to expect even before he and Molyneux launched their assault on Gamma Draconis. Given Gamma's placement in the heavens, Bradley inferred how the Earth's orbital movement would affect the star's position in the eyepiece during the course of the year. Overall, the star should appear to wobble annually along a north-south line: southernmost in December, northernmost in June; in March and September, the star should be situated at the midpoint of its annual wobble. Also, the *rate of change* of the star's position from night to night was expected to vary: gradual change near the endpoints of the wobble, but relatively rapid change near the midpoint. Precisely how far to the south or north Gamma Draconis would swing neither Bradley nor Molyneux could predict; that measure was the long-sought-after parallax.

Installation of the Kew zenith telescope was completed on November 26, 1725. On December 3, Molyneux reclined on a couch underneath the eyepiece and made his first observation of Gamma Draconis when it passed overhead shortly after noon. (Bradley did not arrive in Kew until two weeks later.) Molyneux followed up with three more observations on December 5, 11, and 12. As with Robert Hooke's instrument, the Kew telescope was found to be exquisitely sensitive to environmental influences. The combined body heat of three people standing nearby disturbed the air enough to set the plumb line swaying. Cobwebs had to be regularly cleared from the plumb line, lest they shift the zero-mark from which all measurements were gauged. Nevertheless, the Kew telescope worked remarkably well, with Bradley determining that he could measure a star's position to better than *one arcsecond*. Such precision was unprecedented. Indeed, the two English astronomers had surpassed the capabilities of Tycho's famed Uraniborg quadrants by a factor of 60. It was as though Tycho were trying to read a book through distorting glasses; he'd be able to distinguish light from dark, but the words on the page would be blurred beyond recognition. Bradley and Molyneux, on the other hand, would be able to see the dot on the *i*.

When Bradley arrived in Kew on December 17, Molyneux reported that the position of Gamma Draconis had remained essentially unchanged during his four observations. This appeared to confirm Bradley's prediction that, in December, the star should be hovering around the southernmost point of its annual wobble; no perceptible north-south movement was expected during the next few weeks. Thus, there was no need to observe Gamma Draconis the night of Bradley's arrival. Yet if James Bradley was like every other observational astronomer I know, he probably couldn't wait to get his hands on the new instrument. As he later recalled, "it was chiefly therefore curiosity that tempted me . . . to prepare for observing the star on the 17th."

To both Bradley's and Molyneux's surprise, Gamma Draconis had shifted perceptibly since the observation of December 12. Not only that—the star had moved the "wrong" way: It was now farther *south* than its presupposed southernmost point. And the star continued to slip more and more toward the south as the nights, weeks, and finally months passed. Here was an unmistakable shift in the position of Gamma Draconis, yet one that was wholly contrary to the pattern for a parallax effect. The star's southward creep didn't halt until March 1726, when the star stood a full twenty arcseconds south of its December position. That is, relative to its starting orientation in December, the telescope was now tipped at an angle of twenty arcseconds, less than $^6/_{1,000}$ of a degree; the eyepiece now sat $^3/_{100}$ of an inch sideways from where it had initially been. As March proceeded, the star turned around and began to move northward. In June, it passed its former position of early December, then continued to slide northward until it reached a maximum deviation of twenty arcseconds in September. It turned around once again, this time heading south. Finally in December, it resumed its position of a year earlier.

Bradley and Molyneux continued their observations of Gamma Draconis throughout much of 1727, accumulating a total of eighty position measurements over two years. (Contrast that to Robert Hooke's four observations over four months.) Gamma Draconis was indeed wobbling by a "colossal" forty arcseconds a year. And, given its 365-day cycle, the wobble was undoubtedly linked to the Earth's orbital motion around the Sun. Yet the timing of the wobble was contrary to what was expected for a parallax shift: The star's maximum southward deviation occurred in

March, not in December; its maximum northward deviation occurred in September, not in June. Gamma Draconis was always three months "late." Something was awry, either with the telescope or with the universe. Bradley and Molyneux decided to check the telescope first.

The two astronomers inspected every working part of their instrument and convinced themselves that it was indeed working properly. Molyneux noted that even following "a rainy, blowing tempestuous night" and, later, after "a violent and very unusual hurricane, such as hath not been known in many years," the telescope's alignment remained unaltered. Bradley and Molyneux wondered whether the Earth itself might be wobbling once a year; perhaps Gamma Draconis remains fixed in space, while the supposedly stationary plumb line against which it was measured oscillates. Like passengers who sight a distant lighthouse from the deck of a gently rolling boat, the astronomers might have believed that the faraway, starry beacon swayed, when in fact it was the very deck upon which they stood. However, subsequent observations of a fainter second star ruled out any yearly tottering of our planet. Bradley and Molyneux also considered whether the mistimed wobble might stem from some freakish atmospheric refraction, but here, too, they found evidence lacking.

Until the confounding wobble was explained, every measured star position, every purported stellar parallax, was suspect. With Molyneux suddenly called to service in the British Admiralty, Bradley had to pursue the riddle of Gamma Draconis alone. His first step was to find out whether the strange wobble afflicted other stars outside the restrictive viewing limits of the Kew telescope. From his own modest savings, Bradley commissioned George Graham to make a second telescope, only half the length of Molyneux's twenty-four-foot instrument and with a wider field of view. In August 1727, Bradley returned to the house of his late uncle, James Pound, in Wanstead. With his aunt's permission, he sawed holes in the roof and the floor and mounted the new telescope vertically on the chimney. The house had but one story and the roof sloped low where the chimney poked through it. As a result, the telescope's twelve-foot-long tube placed the eyepiece and the observer's couch below the main floor. James Bradley would be viewing the heavens from the coal cellar.

The new telescope could swing through a twelve-degree arc, 100 times the range of Molyneux's instrument, giving it access to about 200 relatively bright stars, instead of the former handful.

The Wanstead telescope was even more precise than the instrument at Kew: Each full turn of its micrometer screw shifted the eyepiece-end of the tube a mere $^2/_{100}$ of an inch; $^1/_{80}$ of a turn tilted the telescope through a half-arcsecond angle. Despite the telescope's impressive design, Bradley tested the Wanstead instrument "almost to destruction before he would rely on it," in the process declaring a new standard of astronomical practice. Then, over the course of a year, he gathered enough data to conclude that the peculiar wobble was not unique to Gamma Draconis; it occurred in every star he observed. He just couldn't explain why.

Samuel Molyneux had intended to work in tandem with Bradley, continuing to monitor Gamma Draconis from Kew while Bradley studied other stars from Wanstead. Sadly, Molyneux died suddenly in April 1728 at age thirty-nine. His house was sold and eventually dismantled. The Kew telescope was lost. Had Molyneux lived only a few more months, Bradley would surely have trooped over to his house to deliver exciting news. He'd had an epiphany. He knew the meaning of the odd stellar movements they had discovered.

The three princes of Serendip, from the ancient Persian fairy tale, "were always making discoveries, by accidents and sagacity, of things which they were not in quest of." So wrote English author Horace Walpole to his friend, diplomat Sir Horace Mann, in 1754. Walpole named this powerful confluence of happenstance and receptive mind *serendipity*. Serendipity has long been associated with scientific advancement, from the ancient flash of insight by Archimedes—*Eureka!*—to Isaac Newton's falling apple to the modern-era discoveries of Velcro and the microwave oven. Serendipity also makes for a good story.

James Bradley was eminently prepared for the serendipity that blessed him, having puzzled over his wobbly stars for almost three years and still lacking an explanation. In the early autumn of 1728, the answer struck him like a bolt out of the blue. Frustrated astronomer, pleasure cruise on the Thames, gentle breeze. An unlikely formula for solving a cosmic mystery. But that's what makes serendipity so fascinating—its essential improbability. Thomas Thomson tells the tale of Bradley's encounter with serendipity in his 1812 *History of the Royal Society:*

> At last, when [Bradley] despaired of being able to account
> for the phenomena which he had observed, a satisfactory

explanation of it occurred to him all at once, when he was not in search of it. He accompanied a pleasure party in a sail upon the river Thames. The boat in which they were was provided with a mast, which had a vane at the top of it. It blew a moderate wind, and the party sailed up and down the river for a considerable time. Dr. Bradley remarked, that every time the boat put about [turned], the vane at the top of the boat's mast shifted a little, as if there had been a slight change in the direction of the wind. He observed this three or four times without speaking; at last he mentioned it to the sailors, and expressed his surprise that the wind should shift so regularly every time they put about. The sailors told him that the wind had not shifted, but that the apparent change was owing to the change in the direction of the boat, and assured him that the same thing invariably happened in all cases. This accidental observation led him to conclude, that the phenomenon which had puzzled him so much was owing to the combined motion of light and the earth.

Bradley realized that in his observations, he had failed to fully weigh the impact of an orbiting Earth on the perceived positions of stars. He had sought only the inevitable, if as yet immeasurable, consequence of the Earth's changing vantage point on the heavens: stellar parallax. But in doing so, he had inadvertently neglected a much larger phenomenon: the effect of the Earth's *velocity* on his observations. That the Earth, over time, finds itself in different places alters the perceived positions of stars; that the Earth is continually *moving* alters star positions as well.

James Bradley's new insight arose once he connected the essential elements of his telescopic observations with a seemingly unrelated experience: his outing on the Thames. Let's imagine ourselves standing on deck with Bradley on that autumn day in 1728 when serendipity struck. The sailboat tacks up the Thames from shore to shore, spray rising over the bow, breeze riffling our hair. At some point, Bradley's attention is drawn away from the picturesque shoreline. The boat in which we cruise, the wind that fills the sails, the vane that flutters atop the mast are suddenly stand-ins for a trio of astronomical entities. The moving boat plays the part of the moving Earth; the wind becomes rays of light on their beeline course from a star to our planet; the boat's vane, which nominally indicates the direction of the wind, now represents Bradley's telescope, which nominally indicates the position of the star.

Bradley had assumed that a boat's vane behaves like a weathervane on a house; it will always align with the wind, which blows it directly back. This is true, the sailors assure Bradley, only if the boat is stationary. Once the boat is under way, its own velocity also has a hand in guiding the vane. In essence, the boat's forward movement mimics a breeze that blows from bow to stern. Were the boat to ply the waves at ten miles per hour through still air, we would feel a ten-mile-per-hour breeze blowing past us; that we are moving through the air, and not the other way around, makes no difference to our perception. As a result, the vane's orientation is determined by the combined effect of the actual wind velocity and the boat's velocity. That's why we see the vane's direction change every time the boat turns, even though the wind remains constant.

Now Bradley completes the analogy. Whenever he views a star, he reasons, the orientation of his telescope—the "vane"—is determined by the combined effect of the velocity of light—the "wind"—and the orbital velocity of the Earth—the "boat." Were the Earth stationary, as Ptolemy or Tycho would have it, his telescope would show the star in its true position; the tube would be aligned parallel to the incoming light rays, which travel unimpeded down the tube's length. However, since the Earth moves, the telescope is constantly swept along with the planet while starlight streams down the tube. Therefore, Bradley informs us, to center a star in the eyepiece, he must tip his telescope slightly in the direction that the Earth moves; otherwise the star's light will be swept up by the telescope's inner wall before the light reaches the eyepiece. It's as though the beam of starlight entered his telescope tube obliquely. Whichever direction the Earth is heading at the moment, starlight appears to angle into his telescope from that same direction. Thus, Bradley concludes, he sees the star, not in its true position, but skewed a maximum of twenty arcseconds in the direction of the Earth's motion.

Astronomers dubbed the newly discovered phenomenon *aberration*. To those, like myself, of a nonmaritime persuasion, a more familiar instance of aberration occurs when walking through a rainstorm. Even though the rain might be falling straight down, you must tip your umbrella forward to keep the rain from soaking your legs. As the raindrops fall, you intercept them by virtue of your straight-ahead motion. It is as though you stand still and the drops come down obliquely. Whichever way you move, the rain seems to angle toward you from that direction.

In this analogy, you represent the moving Earth, the rain becomes starlight, and your tipped umbrella is the tilted telescope. The angle at which you must tip your umbrella to block the rain depends on how fast you walk relative to how fast the raindrops fall. By measuring your umbrella's tilt and your own velocity, you could compute the raindrops' velocity. Bradley made the corresponding calculation using the tilt of his telescope and the Earth's orbital velocity. He found that light moves so speedily that it covers the huge distance between the Sun and the Earth in eight minutes and twelve seconds, in close accord with modern-era measurements. Therefore, our supposedly "up-to-the-minute" impression of the Sun is always just over eight minutes old. By extension, if we adopt Bradley's minimum distance estimate for Gamma Draconis, light from that star takes at least six years to reach our planet.

Bradley announced the discovery of aberration in 1729 in the form of a letter to Astronomer Royal Edmond Halley. The letter was printed in the Royal Society's *Philosophical Transactions*, under the title, "An Account of a New-Discovered Motion of the Fixed Stars." With their discovery of aberration, Bradley and Molyneux had confirmed the Earth's motion around the Sun as surely as if they had detected a star's parallax. The two astronomers had proven the reality of the Copernican universe.

Bradley continued his observations with the Wanstead instrument, even after his aunt sold the house in 1732. The new owner, Elizabeth Williams, permitted him free access to his now-famous telescope. (The instrument is currently on display at the Royal Greenwich Observatory.) Eighteen years after his aberration announcement, Bradley published the further fruit of his observations from the coal cellar: The Earth's axis, he found, does wobble after all, in a cyclic nodding motion known as *nutation*. Nutation is caused by the Moon's gravitational tug on the Earth's oblate form, an interaction predicted by Isaac Newton. This movement, too, is reflected in the alteration of star positions (by as much as nine arcseconds), along with precession, aberration, and the refraction of starlight by the atmosphere, not to mention the wobbles induced by the telescope itself. The determination of *true* star positions was becoming an ever more complicated affair. If parallax was the proverbial needle, then its companion position-altering effects added up to a formidable haystack.

Having become famous for his discovery of aberration, Bradley found that his lectures at Oxford were suddenly packed, a situation that was duly reflected in a pay raise. By 1732 Bradley could finally afford to live near the university. He moved into a fine house on New College Lane with his aunt and two nephews, but traveled frequently to the house in Wanstead to continue his observations with the zenith telescope. In 1742, Bradley succeeded Edmond Halley as England's Astronomer Royal. He was now director of the Royal Observatory at Greenwich, a post he would hold for the next twenty years. Despite his ascendance, Bradley maintained his propriety: He refused the King's offer of the vicarage of Greenwich, together with its significant stipend, explaining that he could not in good conscience accept a job to which he would devote less than his full measure. It was during his tenure at Greenwich that James Bradley made yet another critical contribution to the quest for stellar parallax: the development of high-precision astronomy.

The Royal Observatory had been founded by England's King Charles II to improve both navigation and its first cousin, positional astronomy. From its inception in 1675 until Bradley's arrival in 1742, the observatory had suffered from chronic underfunding and, of late, from Edmond Halley's lack of attention to its instrumental needs. In short, Bradley found the facility underequipped and in serious disrepair. John Flamsteed, England's first Astronomer Royal (or Astronomical Observator, as he was then called), had acquired the observatory's mediocre original instruments through the generosity of a private patron. In 1688, Flamsteed had purchased better equipment with an inheritance he had received, but these devices, along with most of his papers, were carted off by his widow upon his death in 1719. (The papers were returned decades later; the instruments were lost.) So when Bradley took up his post as third Astronomer Royal, he had at his disposal just two workhorse instruments with which to measure star positions, both purchased during Halley's watch.

These were specialized instruments designed to record the position of a star the instant it crosses the *meridian*. One of the cardinal reference markers of the night sky, the meridian is an imaginary arc that passes from north to south through the zenith and conveniently bisects the sky. Given the Earth's daily rotation, stars east of the meridian are rising, while those west of the meridian are declining. Thus, when a star reaches the meridian, or

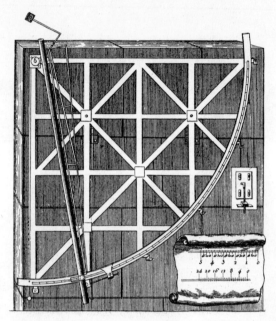

Halley's eight-foot mural quadrant. From Repsold (1908), after the eighteenth-century engraving in Smith's *Opticks*.
Source: *Wolbach Library, Harvard University.*

"transits," it is situated at its maximum altitude for the night, halfway between the eastern and western horizons.

The first piece of meridian equipment Halley had left for his successor was an eight-foot, iron-frame mural quadrant, essentially a telescopic version of Tycho's great mural quadrant in Uraniborg. A mural quadrant is a graduated, metal quarter-circle, mounted flat on a north-south wall (hence, the "mural" qualifier). The sighting telescope is affixed to the quadrant on a pivot, such that the tube is free to swing through an arc from the horizon to the zenith. The telescope is pointed at stars crossing the meridian, and each star's *zenith distance*—its angle from the zenith—is read off the quadrant's scale. The instrument had been designed by George Graham, who had made both Molyneux's and Bradley's zenith telescopes, and triggered a lucrative trade in Graham mural quadrants to observatories on the Continent.

The second measurement device Bradley inherited from Halley was an old, five-foot-long transit telescope that reportedly had once belonged to Robert Hooke. The transit telescope, invented by Danish astronomer Ole Römer in 1689, was mounted crosswise on a horizontal axle, which restricted the telescope's movement to

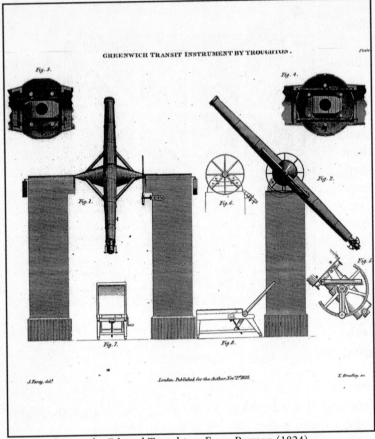

Transit telescope by Edward Troughton. From Pearson (1824).
Source: *Wolbach Library, Harvard University.*

the meridian. An accurate pendulum clock was used to determine the precise instant at which each star crossed the meridian. The timing data was then converted into one of the standard position coordinates used by astronomers.

Together, transit telescope and mural quadrant measurements completely specified a star's celestial coordinates. For greatest accuracy, Bradley measured star positions multiple times, then corrected for effects such as precession, aberration, nutation, atmospheric refraction, and instrumental defects. These "corrected" coordinates he subjected to rigorous statistical analysis before he deemed them acceptable. It was precisely this sort of "unremitting toil" that occupied the bulk of Bradley's life. And by all accounts, he reveled in it. A typical night's work was his observing run at Greenwich for August 8, 1743, when he measured the positions of no fewer than 255 stars. And this while he continued his observing routine from the coal cellar at Wanstead.

Several years into his tenure, Bradley persuaded government officials to provide a £1,000 grant to remedy the Royal Observatory's deficiencies. With the money, Bradley purchased a second mural quadrant, entirely of brass, which he mounted on the opposite side of the wall from Halley's quadrant. Working in tandem, the two quadrants could survey every star that crossed the meridian, Halley's quadrant handling the northern half of the meridian, Bradley's the southern half. Bradley also installed an improved, eight-foot-long transit telescope by John Bird of London, plus his famous zenith telescope from Wanstead. Posthumous studies of Bradley's observations revealed that he indeed measured star positions with an accuracy approaching one arcsecond. This is equivalent to shifting the mural quadrant's sighting telescope along its graduated arc by an amount less than the thickness of a sheet of paper. Feisty Robert Hooke had been correct when he judged open-sight instruments inferior to those with telescopic sights—especially when the telescope lay in the hands of someone as skilled as James Bradley.

After extensive testing to reveal even the slightest mechanical irregularities, from the flexure of the old iron quadrant to the wear of the metal bearings to the slow settling of all the mounting piers, Bradley undertook a long-term program to compile high-accuracy star positions. These positions eventually served as the core of an extensive starry grid against which future generations of astronomers could gauge the motions of planets and comets and ultimately of the stars themselves. Bradley's observations were also the starting point for a breakthrough scientific work in 1818 by a young German astronomer named Friedrich Wilhelm Bessel, who would later catapult to world attention with his own search for stellar parallax.

James Bradley had failed to detect stellar parallax. Nonetheless, he was able to draw a significant conclusion from the negative result: So precise were the Kew and Wanstead telescopes that had the parallax of Gamma Draconis exceeded a mere one arcsecond, it would have been detected. Therefore, Gamma Draconis must be situated at least 400,000 times the Sun's distance from the Earth, or 400,000 astronomical units. This result was no idle guess on Bradley's part, nor was it founded upon controversial assumptions; to believe the vast figure, astronomers only had to trust that

Bradley was able to observe with the claimed one-arcsecond precision. Their trust was buttressed by Bradley's reputation, by the details of his 1729 paper, and by a coincidence of timing. Only a year earlier, Isaac Newton's posthumous *A Treatise of the System of the World* had been published, in which he announced an even larger distance estimate for the bright star Sirius.

With the detection of stellar parallax having proven so problematic, Newton and several contemporaries had developed an alternate means of estimating the distances to the stars, which modern-day astronomer Owen Gingerich aptly summed up in the aphorism, *faintness means farness*. Unlike a parallax observation, which in principle is a definitive measure of a stellar distance, Newton's *photometric* method rests on a very shaky assumption: that all stars are intrinsically identical; that is, the stars in the night sky are merely distant solar "clones." Were Newton's assertion true, then all stars would generate the same amount of light as our Sun. The fainter a star appears in the sky, the farther away it must lie. This simple idea formed the basis of a number of stellar distance studies in the seventeenth and eighteenth centuries. In Newton's photometric method, estimating the distance to a star is equivalent to asking the question: How far would the Sun have to be moved until it resembles one of the specklike stars of the night sky?

The assumption of solar-stellar equivalence implies that even the nearest stars must lie very, very far away. For proof, just compare the blazing appearance of our Sun to the modest glimmer of a star. Precisely how would the Sun's perceived intensity diminish if it were somehow shifted to a greater distance? The intensity of a glowing object appears to fade according to the *square* of its distance from the observer. Imagine two identical light bulbs situated side by side. To the observer, the bulbs will appear equally bright, since both emit the same amount of light and are equally far away. If one of the bulbs is now placed twice as far as the other, it will appear only $1/4$ as bright as its fixed partner; three times as far, $1/9$ as bright; four times as far, $1/16$ as bright; and so on. Suppose the fixed bulb remains ten feet away, while the other bulb is situated at an unknown distance. We can deduce the distance of the second bulb by comparing the relative brightness of the two bulbs. For example, if the second bulb appears $1/25$ as bright as its nearby twin, it is situated five times farther away.

Now replace the nearby bulb with our Sun, and the faraway bulb with a Sunlike star. In principle, we can compute the star's distance by reckoning the relative brightness of the star and the Sun. Thus, Newton and his contemporaries were able to work out rough distances to stars without having to scale the observational Everest of stellar parallax. Even so, the photometric method was difficult to implement; there was no reliable way to compare a star's brightness with the Sun's. After all, the stars are out at night, the Sun during the day. Dutch astronomer Christiaan Huygens tried to maneuver around this difficulty. In his 1698 treatise, *Cosmotheoros,* he described his attempt to estimate the distance to Sirius, the brightest star in the sky. Observing the Sun through a small hole in a disk, Huygens made the hole progressively smaller until the reduced solar image appeared as bright as he recalled Sirius to be. From the size of this "pseudo-Sirius" hole, Huygens computed the relative brightness of Sirius and the Sun, and from this, the distance to Sirius: 27,664 astronomical units.

Newton devised a better technique for gauging the relative brightness of the Sun and Sirius, based on a 1668 idea of astronomer James Gregory. Newton used the planet Saturn as an intermediary between the Sun and Sirius. (Gregory had used Jupiter for the same purpose.) Imagine Saturn as a giant, although imperfect, mirror that reflects a portion of the Sun's light our way. The fraction of the solar light reflected depends on the size of this "mirror-Saturn," its position within the solar system, and its natural reflectivity. Newton mathematically deduced the relative brightness of the Sun and Saturn, then measured the relative brightness of Saturn and Sirius. With Saturn as the unifying link, Newton computed the relative brightness of Sirius and the Sun, and from this, the distance of Sirius from the Earth: 1 million astronomical units. Newton carried out the measurement in 1686, although the result did not appear until 1728, coincidentally a year before Bradley's announcement of the distance to Gamma Draconis. Having perhaps consoled themselves that Sirius's immeasurably small parallax might be atypical, astronomers were confronted with a similar outcome for a second star—and both results from two of the leading lights of astronomy, James Bradley and Isaac Newton. Similar photometric studies of Sirius were completed in 1744 by Philippe Loys de Chésaux, who placed Sirius 240,000 astronomical units from the Earth; in 1760 by J. H. Lambert—500,000 astronomical units; and in 1767 by John Michell—440,000 astronomical units. (The actual distance of Sirius is about 600,000 astronomical units.)

Clearly, by Bradley's time, if not shortly thereafter, astronomers were accustomed to the idea of a vast universe. So they were probably not too shocked by the hard-hitting conclusion of Bradley's 1729 paper on aberration: "[I]t must be granted that the parallax of the fixed stars is much smaller than hath been hitherto supposed by those who have pretended to deduce it from their observations." And not only was stellar parallax exceedingly small, but it also lay buried amidst a host of stellar jogs and jitters that arise from our own planet's atmosphere and movements, as well as from irregularities in the telescope itself. Like murky layers of varnish on a painting, such Earth- and instrument-induced distortions had to be "peeled away" before the underlying true star position could be discerned. Any serious assault on the stellar parallax issue would require a substantial scientific arsenal: an exquisitely balanced, sensitive telescope; almost superhuman perseverance; and the computational wherewithal to tease out the minuscule parallax deviations from those of all the other phenomena that distort a star's position.

James Bradley left sage advice for astronomers who might be tempted to search for stellar parallax. He himself had learned a valuable lesson in his own parallax attempt on Gamma Draconis. That particular star had been targeted for reasons of practicality: It was London's "overhead" star and could be monitored with a relatively easy-to-build zenith telescope. Despite its pivotal role in the discovery of aberration, Gamma Draconis proved to be a poor parallax candidate. It was simply too far away; its parallax was too minute to be detected, even by the most sensitive instrument of the day.

Bradley warned astronomers to be selective in their choice of parallax targets. What folly to invest years of effort gambling on a random star, when that star might be too remote to display a parallax. Somehow astronomers must determine beforehand which of the thousands of stars in the sky are proximate to our solar system, for these are the only stars likely to exhibit a measurable parallax. In his 1747 paper on nutation, Bradley suggested that astronomers examine stars "of greatest lustre," in keeping with the faintness-means-farness doctrine. In addition to studying isolated bright stars, Bradley advised the study of double-star systems that pair a bright star with a faint star. Here he echoes Galileo's second method for measuring stellar parallax, where the bright member of the double system is presumed to lie relatively near the solar system, whereas the faint member is far away. The bright star's parallactic wobble might stand out against its seemingly fixed "partner."

By Bradley's day, the fundamental motive underlying the quest for parallax had already shifted. No longer was stellar parallax needed to prove the heliocentric arrangement of the heavens. Most astronomers were confirmed Copernicans, and those who had wavered were convinced by Bradley's discovery of aberration. The goal of stellar parallax now was to fix the distances of the stars, and thereby transform the night-sky illusion of a starry celestial vault into a full-blown, three-dimensional cosmos. Thus, the parallax hunt had become an end in itself, which astronomers would continue to undertake precisely because it was the most difficult and longstanding challenge in astronomy. Stellar parallax beckoned astronomers the way an unconquered peak draws the mountaineer or a distant horizon calls to the explorer. Until parallaxes were revealed, astronomers would continue to suffer the gnawing disappointment that one of the most fundamental specifications of the universe had eluded them: the distance to the stars. And like the mountaineer and the explorer, astronomers rarely accept such disappointment for long.

SIR WILLIAM HERSCHEL
at the age of 46
*From a crayon copy of the oil painting by L. T. Abbott
in the National Portrait Gallery*

William Herschel. From a crayon copy of the oil painting by L. T. Abbott in the National Portrait Gallery, London.
Source: *Cambridge University Press.*

10

Double Vision

I'm a great believer in luck. I find the harder I work,
the more I have of it.
—*Stephen Leacock*

Chance is always powerful. Let your hook be always cast.
In the pool where you least expect it, will be a fish.
—*Ovid*

★★★

If *People* magazine had existed in eighteenth-century England,
William Herschel would have been on its cover. Everyone from
the head of the local ladies club to King George III was clamoring
to meet this immigrant from Hanover, Germany, who spoke
impeccable English, composed symphonies, built telescopes, and,
of late, had single-handedly changed the popular conception of
the universe. Since time immemorial, it had been common
knowledge that there were but five planets in the solar system
other than the Earth: Mercury, Venus, Mars, Jupiter, and Saturn.
Observers from antiquity onward, first with the naked eye and
later with telescopes, had followed the movements of the five
celestial wanderers as they threaded their way among the fixed
stars. Astronomers of the eighteenth century were able to predict
with fair accuracy where each of the planets would appear on a
given night. And, thanks to Isaac Newton, they had identified
gravity as the agent that held the planets in their orbits about the
Sun. Six planets in all, well-observed, well-explained.

One hundred seventy-one years had elapsed between
Galileo's first telescopic glimpse of Jupiter in 1610 and that fateful night in 1781 that brought William Herschel to worldwide
attention. In that interval, astronomers had scoured the heavens

with their telescopes but had found nothing to alter the certainty that the Sun's family of planets numbered precisely six. Then along came the unknown Herschel—or Mersthel or Hermstel or Herthel or Horchell, as variously reported—and with a home-made telescope that enlarged the planetary census to seven. When the pale bluish-green disk of the seventh planet, Uranus, slid into view, it heralded a profound change in both Herschel's circum-stances and in telescopic astronomy. Whereas professional telescope makers focused their energies on *precision,* Herschel initiated a personal crusade for *aperture.* With increased telescope aperture comes greater light-gathering capacity, hence the ability to see fainter objects and to look deeper into space. Given the state of technology in Herschel's time, lens-based telescopes—refractors—could not be scaled up in aperture without degrading the clarity of their images; a wider lens meant a thicker lens, with all its concomitant distor-tions. On the other hand, mirror-based telescopes—reflectors—could be enlarged with minimal loss in image quality. Herschel personally brought about the explosive growth in reflector-telescope aperture during the late 1700s: from six inches to four *feet* in just over a decade. His vision has reached its current embodiment in today's telescopic behemoths, some of which are wide enough to swallow a house.

With his discovery of Uranus, William Herschel became the eighteenth century's Galileo: telescope maker, explorer of deep space, friend of the common people. Here was a man who com-bined an athlete's drive with a pilgrim's joy, who was as enrap-tured by the sight of a clear, star-filled sky as by the business of plumbing its depths with a telescope. With his naïve self-assuredness, Herschel made the process of discovery seem almost inevitable. "Sweeping," he dubbed his particular brand of methodical observation, as though his telescope were a whisk broom that routinely gathered up unknown planets, stars, and nebulae.

The Herschel phenomenon took the scientific community by surprise. By all accounts, the professionals welcomed the entry of this newcomer into their midst; only months after his discovery of Uranus, the Royal Society awarded him its highest honor, the Copley medal. Still, some scientists were boggled by—and envious of—the sudden ascension of the self-taught musician-astronomer. I imagine they viewed Herschel, to borrow Auden's phrase, as the "shabby curate who has strayed by mistake into a drawing room full of dukes." Who was this man who operated so successfully outside the scientific epicenter of London; who excelled at two diverse occupations, as though able to conjure up hours beyond the daily twenty-four allotted everyone else; who had spiked an otherwise sober technical paper on the Moon with ramblings about lunar inhabitants; who claimed his telescopes magnified several thousand times when no other in the world could achieve better than several hundred? Yet soon even his critics were forced to accept that William Herschel was anything but the "shabby curate," as he appeared before them time and time again with new discoveries about the universe. They could only admire, grudgingly perhaps, the prodigious string of accomplishments spun out of this astronomical dynamo, for no eye had ever before inspected the heavens with such power as his. More than any other astronomer, Herschel opened the door to the investigation of deep space.

William Herschel was a parallax hunter practically from the moment he first pointed a telescope to the heavens. Indeed, the much-heralded discovery of Uranus was incidental to his primary goal at the time: to map the distribution of the Galaxy's stars. "A knowledge of the construction of the heavens," he wrote, "has always been the ultimate object of my observations." Unlike his predecessors, who would have been gratified by the detection of even a single star's parallax, Herschel sought no less than dozens of parallaxes. He aspired to become astronomy's wholesale purveyor of stellar distances. And he believed he had found the means to do it. The latter-day Galileo had reached back to the works of the man himself. He had learned of Galileo's double-star method for measuring stellar parallax: Given a bright star, presumed to lie relatively close to the solar system, in a chance alignment with a faint, faraway star, the bright star's annual parallactic wobble should stand out against its immobile "neighbor." Finding such bright-dim stellar couplings would take many years, for Herschel had "resolved to examine every star in the heavens with the utmost attention"; additional years would be taken up by position measurements and data reduction. But Herschel plunged into his "laborious but delightful research" without hesitation. The parallax project engendered an unwavering, almost fanatical, commitment on Herschel's part. He cast aside well-reasoned objections to the validity of Galileo's double-star method and continued to sweep the skies for double stars, always in the hope that these systems held the key to rendering the third dimension of the near-cosmos. But in the end, William Herschel was forced to confront telescopic evidence that threatened to overthrow the very foundation on which his parallax project stood, evidence that had been amassed by a most formidable critic: himself.

Friedrich Wilhelm Herschel arrived in England from Hanover in 1757 when he was nineteen years old. He immediately Anglicized his name and, except for letters home, wrote and conversed almost exclusively in English. Herschel's father, an oboist in the regimental band of the Hanoverian Guards, had sent his musically minded son across the Channel to escape the latest hostilities with France. Virtually penniless, young Herschel settled in London, where he survived by copying musical scores and by giving violin recitals. Although accomplished on the violin, oboe, harpsichord, and organ, Herschel's ambition was to become a composer. In 1759, he traveled to Yorkshire to assume his first steady post: conductor of Lord Darlington's Durham Militia band (which, Herschel found upon his arrival, consisted of two oboes, two French horns, and a drum). In 1762, he began a four-year stint as organist in Leeds and in Halifax, before leaving to become organist at the Octagon Chapel in Bath. Bath was "the very focus of polite society" in William Herschel's day, with a "parade of blushing damsels and ruffling gallants pictured to our fancy in Miss Austen's novels." The Octagon Chapel was one of several private houses of worship that catered to the religious needs of well-to-do tourists attracted by the city's healing waters. Six fireplaces warmed the opulent chapel, whose doors were closed to lower-class parishioners. From organist, Herschel rose to become Bath's de facto music director. He arranged concerts, composed choral and orchestral music, conducted the choir, and gave music lessons and recitals. (Among those in his choir were several carpenters and joiners, who would later assist him in telescope-making projects.) Following the daily rehearsals and six to eight student lessons, Herschel spent the evenings improving his command of English and teaching himself Italian, Latin, and mathematics. Herschel's situation in Bath proved profitable, and in 1770 he moved into a row house on New King Street. Soon he was joined by his brother Alexander, a talented cellist. And in August 1772, he returned to his native Hanover to fetch his younger sister Caroline.

The death of William Herschel's father in 1767 had stirred no immediate ripples in William's full-throttle life in England. The aftereffects materialized gradually in his growing awareness of his sister Caroline's situation back in Hanover. While his own windows of opportunity were thrown wide open, Caroline's had been shuttered up tight. With the elder Herschel's death, seventeen-year-old Caroline had been left, according to an account by William's granddaughter:

> to the tender mercies of her mother and her brother, Jacob. Neither was intentionally unkind, but her mother was determined to eradicate the foolish notions of improving herself which her father had encouraged and which might lead her to wish to leave home as her brothers had done. A woman's duties, her mother thought, lay in the kitchen and parlor. Caroline was allowed to learn such refinements as washing and darning silk stockings, for Jacob's benefit, and dressmaking for herself and her mother, but nothing more. Jacob's fastidiousness added much to her troubles: "Poor I," she wrote, "got many a whipping for being awkward at supplying the place of a footman or waiter."

Despite the protests of his mother and his brother Jacob, Herschel was adamant that, "Lina," then twenty-two-years old, come with him to Bath and train as a professional singer. In the end, he offered to pay for a household servant to replace his sister. The trip to England was harrowing: six days and six nights in an open mail coach, followed by a storm-tossed passage across the English Channel. Their ship arrived dismasted in Yarmouth, where they were hefted from an open boat on the backs of two sailors and, in Caroline's words, "thrown like balls on the shore." For the next fifty years, they would rarely be apart.

Caroline Herschel was short and slightly built; a dress she wore at William's wedding in 1788 was passed down to a great-niece who was 4-foot-3. There is only one portrait of Caroline as a young woman: a profile entirely in silhouette—symbolic perhaps of the life she would lead in her famous brother's shadow. Caroline's face had reportedly been scarred from childhood smallpox. "I never forgot the caution my dear Father gave me," she recalled later in life, "against all thoughts of marrying, saying as I was neither hansom nor rich, it was not likely that anyone would make me an offer, till perhaps, when far advanced in life, some old man might take me for my good qualities."

Almost immediately upon their arrival in Bath in late August 1772, William began Caroline's voice training, along with lessons in English and mathematics. Within a few years, Caroline was not only running the day-to-day affairs of the Herschel household but was also appearing as featured soprano in William's choral concerts. William never demanded the extreme devotion his sister accorded him, but he was its willing beneficiary. Caroline worked feverishly as housekeeper to shield her brother from all manner of distractions to his music. It had been her idea to perform only where William was conducting, that she would always be available to help him. So in 1778, when Caroline received her first invitation to do a solo performance outside of Bath, she turned it down and summarily gave up her own promising career in favor of assisting her brother's. Decades later, William's son John described his aunt's bond to his father: "She was attached during the 50 years [they were together] as a second self to her Brother, merging her whole existence and individuality in her desire to aid him to the entire extent and absolute devotion of her whole time and powers."

Coincident with his sister's arrival in Bath, William Herschel turned his extracurricular reading from mathematics to science. Caroline recalled that her brother "used to retire to bed with a basin of milk or glass of water, and Smith's *Harmonics* and *Opticks*, Ferguson's *Astronomy*, etc., and so went to sleep buried under his favorite authors." (Ironically, Herschel's interest in science stemmed from childhood French lessons with a technically minded tutor.) Herschel, now thirty-five years old, wrote in his journal on May 10, 1773: "Bought a book of astronomy and one of astronomical tables." The book was *Astronomy*, by James Ferguson, a one-time shepherd who used to lie on his back in the meadow and gauge the separations of stars with a knotted string held up against the sky. Ferguson's book contains twenty-one chapters on the solar system and one on the stars, an indication of the state of astronomical affairs in William Herschel's day. To astronomers, stars were but fixed points against which to measure the motions of planets and comets. These motions, in turn, were employed in ever more stringent tests of Newton's theory of gravity. It was an arcane, highly mathematical effort, widely regarded as the very pinnacle of precise science. To these limited ends, small-aperture quadrants and transit telescopes were wholly sufficient. By virtue of their great distance, stars became the poor stepchildren of astronomy, to be exploited in whatever limited way was deemed profitable. Little attention was paid to their true nature or to the environment they inhabited. In all likelihood, it was this dearth of interest in deep space that inspired William Herschel to adopt it almost exclusively as his own.

Herschel had no intention of becoming an armchair astronomer. "[W]hen I read of the many charming discoveries that had been made by means of the telescope," he wrote, "I was so delighted with the subject that I wished to see the heavens & planets with my own eyes thro' one of those instruments." During the spring of 1773, Herschel put together a crude, forty-power refractor telescope from lenses and a tin tube. This instrument was followed by successively larger ones, up to thirty feet in length, which proved so ungainly that Herschel decided to abandon the refractor design altogether. (Instrument makers were only just learning how to shorten refractor telescopes without compromising image quality; such solutions, however, were expensive.) Herschel rented a small reflector telescope, which performed well enough, yet frustrated him with its meager magnification. Buying a large reflector on a musician's budget was out of the question. Instead, Herschel bought a copy of Robert Smith's 1738 handbook, *A Compleat System of Optics,* and set out to make his own. As luck would have it, an optical hobbyist in Bath offered to sell his entire stock of tools and half-finished mirrors and even to teach Herschel the rudiments of mirror-making. The lessons proved of little value, according to Herschel, the man's "knowledge indeed being very confined."

Herschel settled on the so-called Newtonian design for most of his reflectors. Light enters the upper end of the telescope tube and strikes a concave mirror at the base. This "primary" mirror focuses the light onto a small, flat "secondary" mirror located near the tube's upper end. The secondary mirror is oriented to reflect the light sideways out of the tube and through a magnifying eyepiece. Some of Herschel's larger telescopes dispensed with the secondary mirror; instead, the primary mirror was tilted such that the focused light angled out of the tube to the eyepiece.

In late October 1773, Herschel cast his first disks of solid speculum metal: a whitish compound of copper, tin, and antimony whose surface can be shaped, then polished to a high gloss. (Modern telescope mirrors are made of glass with a coating of reflective metal.) Once cast, each flat speculum disk was abraded by hand into concave form using a disk-shaped tool and powdered emery and rouge. It was laborious, debilitating work. After many attempts, Herschel managed to complete an acceptable five-inch-diameter metal mirror of 5.5-foot focal length, which he mounted in a square wooden tube. He ground his own eyepiece lenses and placed them inside cylinders of cardboard or cocus, a wood used to make oboes. One early biographer mistakenly reported that Herschel made his eyepieces out of coconuts.

It was not long before William Herschel's hobby exploded into an all-consuming passion involving the entire Herschel household. Brother Alexander proved to be mechanically adept— when he could be found. Caroline, on the other hand, was ever-present, if not always enthusiastic. Caroline would read to William and put morsels of food in his mouth while he polished mirrors. Yet in 1773 she complained—silently, of course—that almost every room of the house had been turned into a work-shop: "A cabinet maker making a tube and stands of all descriptions in a handsome furnished drawing-room. Alex putting up a huge turning machine . . . in a bedroom for turning patterns, grinding glasses and turning eye-pieces &c. . . . I was to amuse myself with making the tube of pasteboard against the [eyepiece lenses] arrived from London, for at that time no optician had set-tled at Bath."

A student of Herschel's commented that the room in which he took his music lessons "resembled an astronomer's much more than a musician's, being heaped up with globes, maps, tele-scopes, reflectors, &c., under which his piano was hid, and the violoncello, like a discarded favorite, skulked away in a corner." The same student remarked that once, when the clouds had rolled away from the nighttime sky, Herschel dropped his violin, peered out the window at a star and cried, "There it is at last!" Herschel once admitted, "I was so much attached [to my tele-scopes] that I used frequently to run from the Harpsichord at the Theatre to look at the stars during the time of an act & return to the next Music."

In 1777, Herschel moved the family to a larger house down the block whose backyard provided an unimpeded view of the sky and was large enough to accommodate several telescopes and work sheds. A furnace was installed in the walk-out basement to cast speculum mirrors. The molds that shaped the molten metal were made either of a loam-charcoal composite or, in the case of the larger mirrors, of compressed horse dung. Among Caroline's tasks was to prepare the dung by first pounding "an immense quantity" in a mortar and then sifting it through a fine sieve. It "was an endless piece of work," she noted, "and served me for many hours' exercise."

Like all seat-of-the-pants experiments, there were mishaps along the way. Sometimes the mold cracked from the heat of the molten metal or the mirror itself split while cooling. Once the bottom of the furnace gave way, spilling 540 pounds of fiery metal onto the floor. "Both my Brothers, and the caster and his men were obliged to run out at opposite doors," reported Caroline, who was waiting in the garden, "for the stone flooring (which ought to have been taken up) flew about in all directions as high as the ceiling. My poor Brother fell exhausted by heat and exertion on a heap of brickbatts."

Herschel ultimately became the greatest maker of reflecting telescopes in the eighteenth century. Between 1773 and 1795, he cast, ground, and polished 430 telescope mirrors, the largest of which were a pair, each four feet in diameter. He used no scientific methods to ensure the quality of his instruments; he tried them out on a distant object (the stonework of Windsor Castle became a favorite), then he tried them out on the sky. He always kept in reserve several of his best mirrors with which to compare the images produced by his new ones. At one point, Herschel brought a telescope he'd made to the Royal Observatory in Greenwich. Side-by-side comparisons with the Greenwich instruments showed that his was far superior. "Double stars which they could not see with their instruments I had the pleasure to show them very plainly, and my mechanism is so much approved of that Dr. Maskelyne [England's Astronomer Royal] has already ordered a model to be taken from mine."

In March 1774, Herschel began his lifelong observing journal with descriptions of Saturn's rings and the Orion nebula. Experience suggested that just as he had developed a musician's ear, he might develop an astronomer's eye. "Seeing is in some respects an art that must be learnt," he wrote to a friend. "Many a night I have been practising to see, and it would be strange if one did not acquire a certain dexterity by such constant practice." Indeed, the light of a gigantic, glowing nebula or a 100-billion-star galaxy is reduced by cosmic distance to a faint, hazy wisp in a telescope's eyepiece, easily overlooked by the casual observer. Having "trained" his eye, Herschel commenced the nightly sweeps for double stars; those of unequal brightness he would eventually subject to parallax measurement. In his double-star search, Herschel always used high magnifications; at low magnifications, close double stars can appear as a single star. His telescope's high magnification proved key to what would become his greatest discovery.

Seven-foot, Newtonian-style reflector telescope by William Herschel.
Source: *Royal Astronomical Society Library.*

On Tuesday, March 13, 1781, between 10 and 11 P.M., Herschel
was sweeping the sky with his most effective early telescope: a
seven-foot-long reflector with a 6.2-inch-aperture. Herschel
recollected that while studying the stars, he "perceived one that
appeared visibly larger than the rest." By now a seasoned observer, Herschel knew right away that the object was not a star. A star
maintains its roiling, spiky luminance no matter how much it is
magnified. This object shone steadily and had a perceptible disk.
Herschel continued to observe the object for the next few weeks,
believing it to be a comet. (Comets have no perceptible tail when
they are far from the Sun.) Curiously the object's west-to-east
progression across the sky paralleled those of the planets. And
examining the object at very high magnification, Herschel found
its disk to be perfectly round with a sharp border, unlike a
comet's, which appears nebulous. Herschel notified Neville
Maskelyne, Astronomer Royal, in Greenwich, who observed the
object and responded on April 23. In Maskelyne's opinion, this
was no comet, but a new planet.

"It has generally been supposed," Herschel wrote about his discovery of Uranus, "that it was a lucky accident that brought this star to my view; this is an evident mistake. In the regular manner I examined every star of the heavens, not only of that magnitude [brightness] but many far inferior, it was that night *its turn* to be discovered. I had gradually perused the great Volume of the Author of Nature & was now come to the page which contained a seventh planet. Had business prevented me that evening, I must have found it the next."

Soon congratulatory letters began to arrive in Bath. Distinguished astronomers appeared at Herschel's door. By the end of 1781, Herschel had been elected a fellow of the Royal Society and received its Copley medal for scientific achievement. The unassuming Herschel was taken aback by all the attention. "Among opticians and astronomers," he confided to Caroline, "nothing now is talked of but *what they call* my Great discoveries. Alas! this shows how far they are behind, when such trifles as I have seen and done are called *great*."

In May 1782, Herschel was called to Windsor for a private audience with King George III. He arrived in July with his seven-foot telescope to show the royal family Jupiter and Saturn, the first of many such visits. The king granted Herschel an annual salary of £200 if he would pursue astronomy full time. Herschel's close friend and advisor, William Watson, remarked, "Never bought Monarch honour so cheap!" Although almost as much as the Astronomer Royal's salary, the sum was but half his musician's income in Bath; the difference, Herschel figured, could be more than made up in sales of his telescopes. Caroline was less optimistic. (As it happened, telescope making did prove to be a lucrative sideline business.)

By month's end, Herschel closed out his affairs in Bath, bid farewell to his brother Alexander, and moved with Caroline to a dilapidated house in Datchet, along the marshy banks of the Thames. "I employed myself so intirely [*sic*] in astronomical observations," he wrote from Datchet, "as not to miss a single hour of star-light weather, for which I used either to watch myself or to keep up somebody to watch; and my leisure hours in the day time were spent in preparing and improving telescopes." In 1786, having grown frustrated with four years of Datchet's mist-laden night air and backyard floods, Herschel moved to drier accommodations in Slough. This house, like the one in Datchet, lay within sight of Windsor Castle, as Herschel was frequently called to "entertain" the royal family and visiting dignitaries with his telescope.

In 1783, Herschel completed his largest telescope to date and the primary instrument he would use to sweep the heavens for the next three decades. Nineteen inches in diameter and twenty feet long, the telescope was suspended within a freely rotating, triangular wooden frame. Assistants turned the frame and operated the hand crank that raised or lowered the tube. Herschel dispensed with the usual secondary mirror that deflected light out of the tube; instead he canted the main mirror such that it focused light back to a point near the periphery of the telescope's opening, and here he placed his eyepiece. Celestial images appeared brighter because no light was absorbed in a second mirror reflection. *Front view,* he called his design. Herschel perched on a platform near the eyepiece, some fifteen feet above the ground, while Caroline sat inside the house near an open window, writing down celestial coordinates and descriptions that her brother called out to her. This arrangement had its purpose: The lamplight necessary for Herschel to record his own observations would have spoiled his eyes' accommodation to the dark, or what astronomers term "night vision."

Astronomy, as practiced by the Herschels, was a perilous business, according to Caroline: "I could give a pretty long list of accidents of which my Brother as well as myself narrowly escaped of proving fatal; for observing with such large machineries, when all round is in darkness, is not unattended with danger." Caroline once impaled her leg on a metal hook hidden underneath a blanket of snow. Another night, the entire mounting frame of the twenty-foot telescope collapsed moments after Herschel had climbed down. The astronomer Giuseppe Piazzi, visiting from Palermo, broke his leg in an encounter with a protruding bar.

Herschel's move to Slough in 1786 was triggered not only by his desire for a drier setting but also by an immediate need for more open space. The one-acre garden next to his new house was big enough to accommodate his latest project: a forty-foot instrument with a mirror four feet in diameter. Never before had a telescope of such scale been attempted. Yet King George had been sufficiently impressed by Herschel's telescope-making genius that he granted £4,000 for its construction. Caroline recalled in her autobiography:

William Herschel's forty-foot reflector telescope.
Source: *Owen Gingerich.*

*The gardens and workrooms were swarming with labourers
and workmen, smiths and carpenters going to and fro between
the forge and the 40-feet machinery; and I ought not to forget
that there is not one screw bolt about the whole apparatus but
was fixed under the immediate eye of my Brother. I have seen
him lie stretched many an hour in a burning sun, across the
top beam, when the iron work for the various motions was
fixed. At one time no less than 24 men (12 and 12 relieving
one another) kept polishing, day and night, and my Brother
of course never leaving them all the while, taking his food
without allowing himself time to sit down to table.*

While the one-ton mirror was being prepared in London, the
telescope's nearly five-foot-wide sheet-iron tube sat like an open-air

tunnel in Herschel's yard. The immense cylinder drew the curious, who enjoyed ducking their heads and tramping along its length. King George himself paid a visit on August 17, 1787, with the Archbishop of Canterbury in tow. Part way through the iron shaft, the Archbishop stumbled. King George offered his hand and said, "Come, my Lord Bishop, I will show you the way to Heaven."

Recalling a social tea in 1786 at which she met William Herschel, novelist Fanny Burney remarked, "He seems a man without a wish that has its object in the terrestrial globe." In fact, Herschel was about to embark upon an uncharacteristically earthbound venture: wooing the wealthy widow Mary Pitt, from nearby Upton. In this, he was no less successful than in his heavenly affairs; the two were married on May 8, 1788. If Mary had not already fathomed her husband's devotion to his work, she was soon to learn: Their honeymoon was spent at the house in Slough where Herschel continued to supervise construction of the forty-foot telescope. Six weeks later, the couple settled in Mary's house in Upton, and Herschel "commuted" to work.

Caroline Herschel remained at the house in Slough, but in an apartment above the workshops; the house itself was reserved as a second home for William and Mary. The marriage weighed heavily on Caroline, who later destroyed every reference to this period in her journal. With time, however, she found Mary Herschel to be good-natured, intelligent, and as devoted to William as William was to his work. By the time son John was born in 1792, the "English lady-wife . . . won the entire affection of the tough little German sister."

Herschel completed the forty-foot telescope in 1789, three years after its inception. Decades later, American writer Oliver Wendell Holmes (father of the Supreme Court justice) recalled his first glimpse of the instrument during a visit to England: "It was a mighty bewilderment of slanted masts, spars and ladders and ropes from the midst of which a vast tube, looking as if it might be a piece of ordnance such as the revolted angels battered the walls of Heaven with, according to Milton, lifted its mighty muzzle defiantly towards the sky."

The completion of the forty-foot telescope marked another milestone: Caroline Herschel, having discovered a comet in 1786—the first of eight during her lifetime—and now an astronomer in her own right, was granted a £50 annual salary by the King. She recalled:

[N188]   [In] October I received £12,10, being the first quarterly
           payment of my salary; and the first money in all my lifetime
           I ever thought myself at liberty to spend to my own liking. . . .
           A great uneasiness was by this means removed from my mind,
           for though I had generally (and especially during the last busy
           6 years) been the keeper, almost, of my Brother's purse, with a
           charge to provide for my personal wants with only annexing in
           my accounts the memorandum: "for Car." to the sums so laid
           out, they did, when cast up, hardly amount to 7 or 8 pounds
           per year since the time we had left Bath.

Despite its size, the forty-foot telescope never lived up to its
potential. Prone to tarnish, its mirror required frequent repolish-
ing. Even when freshly polished, the mirror warped under its own
weight, distorting the image in the eyepiece. And preparing the
forty-foot for a night's observations took hours. In the end,
Herschel returned to his mainstay, the twenty-foot telescope.
Nevertheless, the forty-foot's tremendous light-gathering capacity proved its worth on the first full night of use: clearly visible
around Saturn was a previously unseen satellite, later dubbed
Enceladus. The unpretentious Herschel dashed off a note to
Joseph Banks, editor of the Royal Society's *Philosophical
Transactions:* "Would you think it proper, as my paper on
Nebulae is now printing, and I believe one of the last things in the
volume, to add at the bottom of it?—*P.S. Saturn has six satellites.
40 feet reflector. Wm. Herschel.*" Banks complied, and an expanded
version of the postscript was printed. By the time the announcement about Enceladus was issued, Herschel had found a seventh
Saturnian satellite, Mimas. "[If] satellites will come in the way of
my 40-feet reflector," he explained to Banks, "it is a little hard to
resist discovering them."

Herschel's telescopes were simply the biggest around and
were capable of looking deeper into space than any other telescopes in existence. With these instruments, Herschel amassed a
lifelong record of astronomical discovery worthy of an entire
research institute, much less one man. A partial list compiled by
his son John in 1825 includes discovery of Uranus and two of its
satellites, Titania and Oberon; discovery of two Saturnian satellites, Enceladus and Mimas, plus the first measurement of the
rotation period of Saturn's rings; confirmation of the gaseous
nature of the Sun's surface; measurement of the heights of lunar
mountains; discovery of nearly 1,000 double stars and more than

2,000 nebulae and star clusters; resolution of the entire Milky Way into stars; discovery of infrared light; and determination of the solar system's movement through space.

Herschel also employed the twenty-foot telescope for "star gauging," a time-intensive process by which he hoped to determine the shape of the Galaxy. Star gauging—or "gaging," as Herschel spelled it—is rooted in the "faintness means farness" principle. Newton and his contemporaries had applied the rule only to a handful of bright stars; Herschel practiced it on many thousands of stars. Herschel counted stars in 3,400 different sections of sky, then broke down each tally into classes according to brightness. The fainter classes, he assumed, lay farther from the Earth. In this way, he was able to render the Galaxy in three dimensions. The result resembled a ragged disk with chinks of various sizes in its perimeter. The entire star-gauging project rested on one assumption: All stars are identical in their light output. If stars are not intrinsically the same, the results become ambiguous; there is no way to distinguish between a faraway star that appears faint because of its remoteness and a "nearby" star that appears faint because of its weak emission.

Beyond star-gauging's controversial footing, Herschel was unsatisfied with another aspect of the procedure: It yielded only *relative* distances of stars. That is, the shape of the Galaxy was revealed but not its true extent. There remained only one way to measure the distances of stars absolutely: stellar parallax.

"To find the distance of the fixed stars," William Herschel told members of England's Royal Society on December 6, 1781, "has been a problem which many eminent astronomers have attempted to solve; but about which, after all, we remain in a great measure still in the dark." His explanation for this failure echoed that of the ancient Greek astronomer Aristarchus, who initiated the quest for stellar parallax some 2,000 years earlier: "[T]he whole diameter of the annual orbit of the earth is a mere point when compared to the immense distance of the stars." Astronomers had been utterly defeated by this Aristarchean reality and by the attendant minuteness of stellar parallax. Fifty years before Herschel turned his attention to the parallax problem, James Bradley had come up empty-handed in his own attempt on the parallax of Gamma Draconis. Bradley had concluded that parallactic shifts are minuscule, not only in an absolute sense but also compared to the telescopic and environmental disturbances that inevitably accompany their observation; the parallax hunter's only route to success, in Bradley's opinion, lay in the close scrutiny of double stars: Galileo's method. Just nine months after his historic discovery of Uranus, William Herschel announced his intention to take up Bradley's challenge, "so as to be able at last to say, not only how much the annual parallax *is not*, but how much it really *is*."

Herschel planned a three-stage assault on the parallax problem. First he would "sweep" the northern hemisphere sky in its entirety and catalogue every double star that crossed his field of view. Next, he would select from the thousands of newly discovered double stars only those pairs with widely divergent brightness. Finally, he would perform the requisite long-term position measurements of each selected pair. Herschel's prime quarry was very narrowly spaced double stars, whose members were separated by no more than five arcseconds. In such "tight" doubles, the distorting influences of atmospheric refraction, aberration, and the various earthly jigs and wobbles affect the member stars equally and therefore can be ignored. Any annual wobble of the bright star relative to the dim one, Herschel reasoned, can thus be attributed unequivocally to parallax.

Herschel understood that his own telescopes were among the world's best, especially at high magnifications; a star system that revealed itself to be double in his instruments might appear as a merged blob of light to other astronomers. He reassured his colleagues that he "guarded against optical delusions" and that the closely spaced doubles only he was able to discern did indeed exist. Further, an observer should "not condemn his instrument or his eye if he should not be successful in distinguishing them." In January 1782, Herschel published his first double-star catalogue, listing 269 systems he had encountered in a two-year sweep of the sky; successive sweeps eventually raised that number to almost 1,000. Using an eyepiece micrometer—a pair of movable, parallel wires within the telescope's view field—he recorded the relative orientation of each star pair and the separation between the two members.

It wasn't long before Herschel's parallax project received its first jolt. In April 1782, Astronomer Royal Neville Maskelyne directed Herschel's attention to a 1767 paper by Cambridge geologist and astronomer John Michell, entitled "An Inquiry into the probable Parallax, and Magnitude of the fixed Stars." (Michell is also acknowledged to have described the concept of the black hole in 1784.) Through statistical analysis, Michell had demonstrated that a random distribution of stars in space would produce relatively few chance alignments as viewed from Earth; that is, only a tiny fraction of double stars are the result of an alignment of a nearby star with a faraway star. In Michell's words, it is "highly probable in particular, and next to a certainty in general, that such double stars, &c. as appear to consist of two or more stars placed very near together, do really consist of stars placed near together."

If most double stars are true pairs—"binary" stars—they cannot be used to measure parallax the way Galileo had proposed. Herschel, like Galileo, had assumed that all stars are essentially clones of the Sun and that any difference in the apparent brightness of two stars stems from a difference in distance. It is this supposed difference in distance upon which Galileo's method relies. Michell, on the other hand, held that the pairing of a bright with a dim star reflects a difference in stellar light output, not distance: The bright star simply produces more light than its partner. If true, the entire foundation of Herschel's parallax project would crumble.

In 1783, Michell wrote an open letter to the Royal Society in which he revealed the shaky basis of Herschel's parallax effort. He also predicted that within a few years there would be conclusive observational evidence that "some of the great number of double, triple, stars, etc., which have been observed by Mr. Herschel, are systems of bodies revolving about each other." Herschel countered that it was too soon to make a final judgment about the nature of double stars. And until such time as there was proof of Michell's assertion, he would proceed with his plan to measure stellar parallax using Galileo's method.

Other work intervened, and although Herschel continued to sweep for double stars, he effectively put the parallax project on hold until 1802. That's when the second jolt came. Herschel decided to reinspect several double-star systems he had catalogued some twenty-five years earlier. In almost every case, he found that the relative placement of the member stars had shifted over time. Their manner of movement was unmistakable: The stars were orbiting one another. John Michell had been right: Most double stars are true pairs, not chance alignments, and here was the observational evidence that proved it—evidence that negated Herschel's entire plan to measure stellar parallax. Evidence that would be put forth by Herschel himself.

During the summer of 1802, while a despondent Ludwig van Beethoven retreated to the village of Heiligenstadt to contemplate his worsening deafness, William Herschel composed the words that would end his longtime dream of measuring the cosmos. "I shall soon communicate," Herschel announced to the Royal Society on July 1, 1802, "a series of observations made on double stars, whereby it will be seen, that many of them have actually changed their situation with regard to each other, in a progressive course, denoting a periodical revolution round each other." With the publication in 1803 and 1804 of these observations, detailing the orbital movements of several dozen binary systems, William Herschel's parallax quest came to a close.

Herschel's revelations about the true nature of double stars dealt a mortal blow to another longstanding project of his: star gauging. The very existence of bright-dim binary stars proves that stars are not all alike, as Herschel had assumed, but exhibit a wide range of light output. (By modern measure, stellar light emission varies anywhere from less than $1/1{,}000$ to 1 million times the Sun's luminosity.) Therefore, stars are not uniform beacons whose brightness reveals their relative placement in space. Even in the face of his own damning evidence, Herschel continued in his later years to gauge the stars in a vain attempt to delineate the borders of the Galaxy.

However personally distasteful his own conclusions about the true nature of double stars, Herschel had in fact demonstrated something of profound significance: the universality of Newton's laws of gravity and motion. Newtonian gravity not only plucked apples from tree limbs on Earth and held the planets fast about the Sun, but, far beyond the confines of our solar system, also enfolded pairs of stars in its invisible embrace. Decades later, astronomers would use Newton's equations to compute the masses of Herschel's binaries. Having failed in his parallax effort, Herschel had given astronomers the means to "weigh" the stars. Stellar mass is one of the crucial ingredients in modern-day theories of how stars create their energy. The observational seed planted by Herschel in the early nineteenth century was to blossom in the stellar astrophysics of the twentieth century.

"Lina, there is a great comet," began William Herschel's final note to his sister. "I want you to assist me. Come to dine and spend the day here. If you can come soon after one o'clock we shall have time to prepare maps and telescopes." Caroline Herschel dated the note—July 4, 1819—and wrote on it, "I keep this as a relic! every line now traced by the hand of my dear Brother becomes a treasure to me." Her brother was by this time eighty-one, and her journal was increasingly given over to concerns about his declining health. The journal ceases in October 1821, Caroline unable to chronicle her private grief any further. She was at William's bedside when he died on August 25, 1822.

Less than two months later, Caroline stepped aboard a steamer packet, bound for her native Hanover, leaving behind the country she had called home for fifty years. She moved in with her brother Dietrich's family for three years and upon his death lived another twenty-three years alone. In Hanover, Caroline was a reluctant celebrity, hosting dignitaries from all over the continent. In 1828, the Royal Astronomical Society awarded her a gold medal for the catalogue she had recently assembled from her brother's observations. Seven years later, she became the first woman to receive honorary membership in the Society; the Royal Irish Academy followed suit in 1838. And in 1846 the Prussian king presented her with his country's gold medal for science. Yet honors mattered little to Caroline Herschel, as did accolades from scientists and kings. "You will see what a solitary and useless life I have led these 17 years all owing to not finding Hanover, nor anyone in it, like what I left, when the best of brothers took me with him to England in August, 1772."

John Herschel, trained by his father William, became an eminent astronomer in his own right, who received his share of medals and eventually rose to the presidency of the Royal Astronomical Society and the British Association for the Advancement of Science. By the time he visited his Aunt Caroline in Hanover in July 1832, he was Sir John Herschel. He told Caroline of his intention to gather the old twenty-foot telescope and journey with his wife Margaret to the Cape of Good Hope, and there to sweep the southern hemisphere skies as his father had done in the north. "Ja!" Caroline responded to the news, "if I was but thirty or forty years younger, and could go too!"

John and Margaret Herschel left England for their five-year stay at the Cape on November 13, 1833. Only months earlier, an acquaintance of John's, astronomer Thomas Henderson, had been sailing northward from the Cape, back to his native Scotland.

Henderson had completed a miserable year at the Royal Observatory in South Africa. His health compromised and his upper-crust sensibilities shaken, he had become desperate to return to the comforts of home. At the time, Henderson hadn't fully realized just how fruitful his mission at the Cape had been, for hidden like a stowaway among his mass of observational data was the first evidence of stellar parallax.

⋆ Part 3

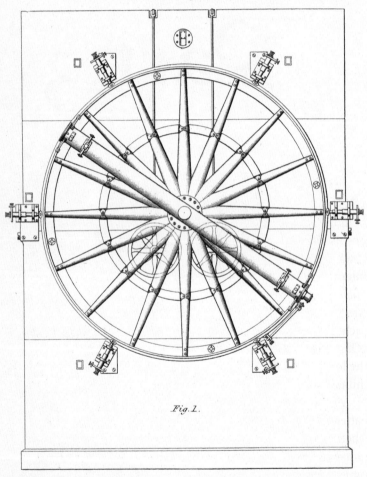

Front Elevation of the Circle.

Fig. 1.

Greenwich mural circle by Edward Troughton, similar to one at the Royal Cape Observatory, South Africa. From Pearson (1824).
Source: *Wolbach Library, Harvard University.*

11

Dismal Swamp

I will tell you about my residence in the Dismal Swamp
among slaves and savages . . .
—Thomas Henderson at the Cape of Good Hope,
in an 1833 letter to Thomas Maclear

Fortune brings in some boats that are not steered.
—William Shakespeare, *Cymbeline*

★ ★ ★

Nine miles south of Arecibo, Puerto Rico, in a rolling landscape
of hillocks and hollows, lies one of the world's more unlikely out-
posts for astronomical research: the Arecibo Radio Observatory.
It was here that I spent the summer of 1972 as a college-student
intern. Having landed at San Juan airport, I rode in a crammed
publico—a cross between a taxi and a cattle car—to the city of
Arecibo, where I'd planned to hitchhike out to the observatory. To
each driver who offered me a ride I said in faux-Spanish,
"*Observatorio.*" The observatory, I found out later, was known to
the locals as *El Radar,* supposedly a reference to some top-secret
military project being carried out there.

About an hour later, I was sitting in the passenger seat of a
battered VW Bug, heading up the access road that winds inland,
when all of a sudden I saw out the window a gleaming, 1,000-
foot-wide, wire-mesh bowl nestled among the hills. Suspended by
cables above it was a giant steel-girder triangle, whose underside
sported an assortment of spearlike antennae. The cables hung
from three towering concrete pylons that rose from hilltops
around the bowl's periphery. All with the lush Puerto Rican landscape as a backdrop. The whole structure looked as out of place as
a battleship in a ballpark.

Unlike a conventional telescope, the Arecibo instrument detects cosmic radio waves instead of visible light. Since its completion in 1963, it has revealed to astronomers a "hidden universe" of celestial objects—pulsars, quasars, supernova remnants—that co-inhabits the universe we see but whose emissions are invisible to the eye. Yet it was a different sort of unseen universe that impressed me even more, one that unveiled itself every evening after the sun went down: the night sky of Puerto Rico. Only once before, while on vacation in the Colorado Rockies, had I witnessed a celestial scene that approached the magnificence of what I saw from Arecibo. The Milky Way looked bigger, richer, more varied, with pools and canyons of darkness trespassing across its luminous extent. Gazing up at the star clouds of Sagittarius, in the direction of the galactic center, I almost reeled backward under the "weight" of the multitude of stars above me.

Situated at a latitude 23 degrees south of my former home in New Jersey, my Puerto Rican viewing site literally expanded my horizons. For the first time, I fully appreciated how for all my years certain reaches of sky had been shunted out of view, eclipsed by the very ground upon which I had grown up. Now joining familiar starry patterns such as the Big Dipper and Cassiopeia were constellations I had never encountered—constellations of the southern hemisphere. And hovering almost due south every night of that long-ago summer, shining like a cosmic lighthouse, was brilliant Alpha Centauri. The third brightest star in the heavens, Alpha Centauri had been worshiped by the ancient Egyptians. And, most significant, it is the solar system's nearest stellar neighbor, whose parallax is larger than that of any other star. In short, a parallax hunter's dream.

The best telescopes of the early nineteenth century—and even a few from the eighteenth century—could have detected the parallax of Alpha Centauri. Yet those telescopes sat in European observatories, too far north to sight the star. Pinning down Alpha Centauri's parallax had to await the establishment of a well-equipped, southern hemisphere observing station. Bound up in that effort are the stories of two astronomers: the Reverend Fearon Fallows, who built the first permanent southern observatory, and his reluctant successor, Thomas Henderson, a proper gentleman from Scotland, who quite inadvertently measured the parallax of Alpha Centauri.

Until late in the twentieth century, when a series of large-aperture telescopes saw "first light" on various mountaintops in South America and Australia, southern hemisphere astronomy had always played poor cousin to its northern hemisphere counterpart. By 1800 William Herschel had already surveyed the whole of the northern sky with his big reflector telescopes. The southern sky, by contrast, had remained a cosmic wilderness; no one had scanned its extent with any instrument larger than a half-inch in diameter. Except for a few hardy souls who had set sail to locales below the equator, the southern hemisphere had been virtually ignored by astronomers. Constellations themselves were up for grabs, with various observers proposing alternative, often overlapping, sets of star patterns. Historian Agnes Clerke summed up the chaotic situation in the south:

> Celestial maps had become "a system of derangement and confusion"—of confusion "worse confounded." New asterisms [star patterns], carved out of old, existed precariously, recognized by some, ignored by others; waste places in the sky had been annexed by encroaching astronomers as standing-ground for [constellations of] their glorified telescopes, quadrants, sextants, clocks. . . . [P]alpable blunders, unsettled discrepancies, anomalies of all imaginable kinds, survive in an inextricable web of arbitrary appellations, until it has come to pass that a star has often as many aliases as an accomplished swindler.

One of the first astronomers to turn his eyes south of the equator was young Edmond Halley, who sailed in November 1676 to the south Atlantic island of St. Helena. In addition to witnessing a transit of Mercury across the face of the Sun, Halley observed a lunar eclipse and a solar eclipse and recorded his impressions of many previously unstudied star clusters and nebulae. He also used a sextant with telescopic sights to measure the positions of 341 southern stars. Referring to Halley's new southern hemisphere star chart, contemporary biographer John Aubrey commented pointedly that "his Majesty was very well pleased with it; but [Halley] received nothing but Prayse." Perhaps mindful of the King's role in securing funds for astronomical research—the King had arranged Halley's free passage to St. Helena—the young astronomer carved out a portion of the southern constellation Argo (the ship) and renamed it Robur Carolinum, or Charles's Oak. For his part, King Charles II directed Oxford to award Halley a degree in 1678, even though Halley had not completed the graduation requirements. From this auspicious beginning, Halley's star rose steadily until he succeeded John Flamsteed as England's Astronomer Royal.

In 1751, French astronomer Nicolas Louis de Lacaille sailed even farther south, bringing a sextant and small telescope to the Cape of Good Hope. There he catalogued 10,035 stars, measured the parallax of the Moon and Mars (in a cooperation with astronomers in Europe), and proposed the adoption of fourteen new southern constellations. "When it is remembered that [Lacaille] did all this work single-handed," astronomer David Gill reminds us, "one is lost in amazement at the energy and capacity of the man. In the course of a single year he laid the foundations of sidereal astronomy in the Southern hemisphere."

Both Halley's and Lacaille's expeditions demonstrated the need for a permanent observing facility below the equator. Handheld or crudely supported instruments simply could not match the power and precision of the permanently mounted telescopes up north. Finally, in 1820, England's Board of Longitude voted to establish the southern hemisphere's first major astronomical observatory, to be located in South Africa at the Cape of Good Hope. That was the easy part. Now the board had to find an astronomer—highly trained, a Cambridge or Oxford man, of course—willing to be posted to a remote colony with few amenities and a decidedly un-English climate. As it happened, the man found them. On February 19, 1820, thirty-one-year-old Reverend Fearon Fallows wrote to his friend, the eminent astronomer John Herschel, about a rumor he had heard: "There is a report here [in Cambridge] that Government has determined to erect an Observatory at the Cape of Good Hope & to appoint an observer with two assistants to manage it. Pray have you heard anything about it as I have some notion, in case of such an event taking place, of offering myself as a Candidate provided the Salary was liberal." Herschel contacted Lord Palmerston at the Admiralty, and the deed was done. Fallows was appointed director of the yet-to-be-built Royal Cape Observatory.

A one-time weaver who used to read mathematics texts at the loom, the Cambridge-educated Fallows was well versed in the theoretical aspects of astronomy, but had scant experience at the telescope. He embarked on a "crash course" in the techniques of astronomical observation, visiting observatories and telescope workshops throughout England, before shipping out to the Cape. He was never to return to his native land.

Fallows sailed into Cape Town's Table Bay in August 1821 with his new wife, Mary Anne, and an observing assistant. The troubles began right away. The civil authorities in Cape Town knew nothing of Fallows's mission. The Admiralty had neglected to make arrangements for housing or for the transport of the ten tons of astronomical gear sitting in the ship's hold. Nor had any money been sent to temporarily mount the instruments while the observatory was being built. Fallows rented a house for himself and his wife and found a warehouse to store the equipment, paying for everything out of his own pocket. Meanwhile, the South African climate began to take its toll; within three months, Fallows was bedridden with sunstroke. He wrote to Herschel again: "Conceive yourself in a far distant land, sent out under the highest patronage for the promotion of the most noble objects, & an apparently insuperable bar to any success presents itself at the very onset. . . . I was nearly overwhelmed with grief and disappointment."

Partially recovered, Fallows tramped all over the countryside to find the best location for the new Royal Cape Observatory. Fallows was a consummate walker, having once traveled fourteen hours on foot from Cambridge to London. He scrutinized every pristine hilltop within forty miles of Cape Town, places with names like Blue-berg, Elephant's Head, and Tiger Hill, and unnamed sites like the "spot near Mr. Coetsey's farmhouse." In January 1822, Fallows communicated to the Admiralty his first choice—Tiger Hill—with a raft of supporting evidence. But by March, he'd abandoned that site for another that promised to be even better, at least from the astronomer's "professionally myopic" viewpoint. (Astronomers routinely park their telescopes on remote mountaintops, eagerly trading comfort and convenience for a shot at the clearest, darkest skies.)

The new site chosen by Fallows was a barren rise referred to by the locals as Slang-kop, about three miles southeast of Cape Town at the confluence of the Liesbeck and Salt rivers. Astronomically speaking, Slang-kop appeared to Fallows as a first-rate location for an observatory. Its horizon was unobstructed, except toward the southwest where Devil's Peak and Table Mountain reared their stony faces. There was a clear line of sight toward the harbor in Table Bay, allowing Fallows to relay time signals in the evening to anchored ships via lantern and shutter. And most important, Slang-kop was sufficiently tall that the Cape's notorious dust-laden winds dropped much of their burden before howling over the summit.

There was, however, a problem with the proposed building site, a problem foreshadowed by its very name: Slang-kop, translated, means *Snake Hill*. And Snake Hill lived up to its billing; nearly 100 snakes dashed from the thistle-covered knoll during groundbreaking. Fallows described how workmen threw them out of the hole by the handful. Jackals also frequented the area, and the occasional hippopotamus could be seen wallowing in the marshes below. Once, years later, Fallows stepped outside to investigate why the observatory's protective shutter was stuck, only to find a leopard lounging on it.

Construction of the new observatory did not start until 1825, following three years of bureaucratic snafus and local opposition. (One resident had argued that the money would be better spent on a breakwater for the harbor.) Even then, the work crawled along, delayed by difficulty transporting building materials and by the miserly attitude of the Admiralty, which held the purse strings. On one occasion, Fallows requested money for a bank of trees to shield the observatory from the wind; the Admiralty countered that it was not about to foot the bill for a beautification project. Fallows, too mild-mannered to protest, suffered repeated setbacks in stoic silence. As one of his biographers observed, "It is difficult to conceive [of] a man of such simplicity of character and such absence of knowledge of the world in the nineteenth century.... He had apparently little capacity for dealing with men ... and it is clear from the official correspondence that his deficiency in this respect, and in business matters generally, produced an unfavourable impression at the Admiralty."

Nevertheless, Fallows proved to be an indefatigable observer, undeterred by heat, wildlife, bureaucrats, or social isolation—an astronomical missionary whose calling was to bring uncharted stars into the fold. As construction on the permanent facility lumbered on, Fallows started observing from his home in Cape Town with a 1.6-inch-aperture portable transit telescope by Dollond. An entry in the Cape Observatory's archives reports that his landlord, van Breda, "one day took offence at Mr. Fallows looking into Breda's backyard from the top of the house, Breda contending that Mr. Fallows came to the Colony for the sole purpose of looking upwards and not downwards." The complaint went to court;

Portable transit telescope used by Fearon Fallows.
Source: *Clive Booth, South African Museum.*

Fallows was evicted. He moved his operations out to Snake Hill and observed the heavens from there, while wild creatures padded by in the shadows. During the day, he ran a school for the children of local farmers; tuition for each day's lesson was a load of topsoil for Snake Hill's hard-packed summit.

The observatory was finally completed in 1828. It was a handsome, neoclassical structure of quarried stone, plaster, and even Burmese teak that had been purchased cheaply from a passing ship. The building was laid out like a letter H, with space for four major instruments: two in the east-west "crossbar" section and one each in the north-south wings. In 1829, Fallows shifted his observing program to the permanently mounted instruments: a ten-foot-long Dollond transit telescope and a supposed replica of

Thomas Bowler's 1834 drawing of the Royal Cape Observatory, with Devil's Peak and Table Mountain in the background.
Source: *South African Astronomical Observatory.*

Edward Troughton's mural circle at Greenwich, pictured at the opening of this chapter. The wall-mounted mural circle is the logical extension of the mural quadrant. Whereas a pair of opposite-facing quadrants is necessary to access the sky from horizon to horizon, the same can be accomplished with a single mural circle.

To his dismay, Fallows learned that the mural circle had slammed into the dock while being hoisted from the ship. He knew that the slightest asymmetry of the central axle might render the instrument virtually useless for precision star-position measurement. This was not something that could be hidden from the astronomical community, even if Fallows had been so inclined; mechanical flaws declare themselves as systematic errors in the data. Once an instrument gains a bad reputation, little can be done to rehabilitate it. (The same can be said for "flawed" observers. Astronomers keep track of such things—discreetly.) Fallows determined that acceptable star positions could still be obtained if he averaged the individual readings of all six of the mural circle's position indicators. This discovery later proved significant to the measurement of Alpha Centauri's parallax.

One couldn't blame Fearon Fallows if, by this point, he were tempted to believe the Fates were toying with him. His first observatory assistant, a drunkard, ran off with the housemaid, also a drunkard. The second, a Catholic chaplain whom Fallows had originally praised to the Admiralty, seduced the maid's seventeen-year-old replacement and was dismissed. The third fell ill and shipped back to England. Desperate, Fallows enlisted the help of his wife, Mary Anne. With her assistance, he ultimately completed nearly 4,000 individual position measurements of southern stars. He also reported to the Admiralty that Mary Anne herself had discovered a comet in the constellation Octans.

In September 1830, circumstances shifted from merely frustrating to truly life-threatening. Fallows and his wife contracted scarlet fever. Although he insisted on continuing his observation program, Fallows eventually grew so weak he had to be carried to the telescope in a blanket. He died on July 25, 1831, at the age of forty-two. At his request, the precision-minded Fallows was buried on the observatory grounds, twelve feet down, due south of the building's main entrance. Mary Anne recovered from her illness and returned to England, carrying her husband's papers and unpublished observations. She remarried in 1835, but bled to death three years later after a medical application of leeches.

Several months after Fallows's untimely death, the Admiralty nominated Thomas Henderson as the second Royal Astronomer at the Cape of Good Hope. A thirty-three-year-old law school graduate and former clerk to the Chief Justice of Scotland's Supreme Court, Henderson had taken up astronomy as a hobby when he was a teenager. He'd learned practical astronomy at Calton Hill Observatory, operated by the private Astronomical Institution of Edinburgh. Mathematically gifted, Henderson burst into the astronomical spotlight during a creative spurt, when twelve of his papers were published in the span of only three years. In particular, his new method to predict eclipses of stars by the Moon (occultations) had been featured in the 1827 edition of the *Nautical Almanac*.

Henderson accepted the Cape post with great reluctance, on the advice of friends, after his professional aspirations had suffered a double blow. Despite being highly recommended, he'd been turned down for the Chair of Practical Astronomy at Edinburgh University, and then passed over for the position of Superintendent of the *Nautical Almanac*. It appeared that the only path to a full-time career in astronomy led through the Cape.

Henderson reached South Africa in April 1832. He hated the place from the start. For years he'd frequented Edinburgh and London, and through his legal dealings on behalf of various wealthy men had grown accustomed to associating with persons of high social standing. Although committed to his astronomical work, he nonetheless valued creature comforts. The Royal Cape Observatory lay just three miles outside of Cape Town, but to Henderson it must have felt like the remotest outpost on Earth. Standing on the portico of the observatory, gazing beyond the black stone that marked his predecessor's grave and down to the marshes below, Henderson surely longed for the comforts of home. He coined his own name for the place, a name that appeared repeatedly in his correspondence: the "Dismal Swamp."

Saddled with a weak constitution and incipient heart disease, Henderson commented candidly and often about the living conditions in the Dismal Swamp, particularly the seemingly endless supply of "insidious, venemous [sic] snakes." To his eventual successor, Thomas Maclear, he wrote, "What would you think if, on putting out your candle to step into bed, you were to find one lurking beside the bed?" Henderson also made his feelings known to the Admiralty. He summed up the deficiencies of the Cape Observatory in this 1833 missive:

> Perhaps I may be pardoned for taking the liberty of recommending to their Lordships' consideration the state of the observatory, which I am afraid would, in the opinion of every British subject who takes an interest in science and regards the honour of his country, be deemed not satisfactory. . . . Its situation upon the verge of an extensive sandy desert, exposed to the utmost violence of the gales which frequently blow, without the least protection from trees or other objects to shelter from the wind or sun, some miles distant from markets, shops, or the inhabitations of persons with whom those belonging to the observatory can associate, the want of good water, and the state of the bulk of the population from whom servants must be taken and other aid applied for, will always prove considerable drawbacks from the comforts of persons sent from England to do the duties of the observatory, and great obstructions to the undisturbed cultivation of the science.

Despite his vocal reservations about the Cape, Henderson was "too honourable a man to accept the emoluments of an office without the most punctilious discharge of his duties." His accomplishments during the year following his arrival were impressive by any measure. He determined the precise latitude and longitude of the observatory, values that still stand today. He measured the parallaxes of the Moon and Mars, and from the latter inferred the Sun's distance. He tracked the paths of Encke's and Biela's comets, recorded eclipses of Jupiter's satellites and occultations of stars by the Moon, and timed the transit of Mercury across the face of the Sun. Henderson also dramatically accelerated the program to chart southern hemisphere stars. For efficiency's sake, he operated the Cape's two instruments in tandem; he himself used the troublesome mural circle, while his assistant, Lieutenant William Meadows, used the Dollond transit. In the end, Henderson produced a catalogue of southern star positions every bit as accurate as those for the northern hemisphere. He was also one of the first British astronomers to objectively assess the accuracy of his observations according to the mathematical theory of statistics.

A year into his tenure at the Cape, Henderson's patience ran out. He'd lost the capacity to deal with the daily struggles of life at the frontier and the apparent indifference of the Admiralty to his personal and professional needs. He missed his comfortable gentleman's existence back home. As one biographer succinctly put it, "Henderson was rather the refined observer than the pioneer." On April 27, 1833, Henderson resigned his position as Royal Cape Astronomer on the pretext of a dispute with local authorities about the observatory's sanitary facilities. Within a month, he had packed his belongings and boarded a vessel to England. He was not about to wait for the Admiralty to appoint a successor.

Having sailed out of Cape Town, Henderson must have had a lot on his mind. The resignation had been a desperate act, especially for someone in his financial circumstances. No job awaited him at home; he'd have to make ends meet on the £100 annual pension from his Supreme Court clerkship. He didn't know at the time that his service in the Dismal Swamp would shortly pay off—in just over a year, he would be appointed not only Regius Professor of Astronomy at Edinburgh University but also Scotland's first Astronomer Royal.

And there was another matter to be dealt with. Henderson carried in his observing logs an astronomical bombshell: nineteen precise measurements of Alpha Centauri's position. Only weeks earlier, the star had been just one of many on his observing list, another line of data in his soon-to-be-published catalogue. Not anymore. Henderson had received word from a colleague on St. Helena that Alpha Centauri was remarkable for more than its brilliance. Alpha Centauri was moving. And moving fast.

A century earlier, Edmond Halley had launched the study of stellar motions by comparing the modern positions of several prominent stars against those recorded by the ancient astronomer Hipparchus. To his surprise, he found that Sirius, Aldebaran, and Arcturus had shifted since antiquity. In other words, at least some of the so-called fixed stars are not fixed at all. Astronomers call such long-term stellar movements "proper motion." Proper motion is our impression of a star's actual movement through space. Assuming that all stars drift through space at roughly the same speed, those closest to Earth should display the largest proper motion. By analogy, a hawk flying just above the treetops appears to streak by quickly, whereas an equally speedy hawk flying at high altitude seems to inch across the sky.

Alpha Centauri's proper motion, Henderson had only recently learned, was 3.6 arcseconds per year, equivalent to a sideways movement of 1.5 millimeters viewed over the length of a football field. A minuscule shift by ordinary standards, but a whopping figure compared to that of most other stars. Henderson realized that Alpha Centauri, in view of its large proper motion, was very likely a nearby star. In his nineteen measurements of the star's position lay the strong possibility of at least an indication of the telltale wobble of parallax. Henderson was also aware that nineteen observations were insufficient to prove *beyond question* the existence of Alpha Centauri's parallax. Had he found out sooner about the star's large proper motion and likely proximity, he would have made a greater number of position measurements. The more measurements, the more accurate—and thus the more believable—the deduced parallax. But now it was too late. He was sailing out of Table Bay, bound for home up north, where Alpha Centauri would be forever out of sight.

Henderson had an important decision to make. The mural circle telescope he had inherited from his predecessor, Fallows, was known to be defective; any parallax deduced from its observations would be viewed with suspicion by the astronomical community. Astronomers had read too many confident reports of stellar parallax that, upon further scrutiny, proved to be questionable or even downright false. In the quarter-century between 1800 and 1825, more papers about stellar parallax had been published than in the century-and-a-half prior to 1800. Results drawn from mural circles in particular had become a favorite target of critics. Henderson faced a dilemma. He could deduce Alpha Centauri's parallax using the nineteen observations, publish the result, and stake his claim as the first to have measured the long-sought-after stellar parallax. Or he could wait while his assistant, Lieutenant Meadows, completed a set of corroborating observations using the Cape's transit telescope. In the first case, Henderson risked being viewed by his colleagues as the next in a long line of hack observers vying for glory. In the second case, he chanced losing the parallax "race" altogether.

Every prior stellar parallax claim had been rejected by the astronomical community, regardless of the claimant's reputation or best intentions. Like jurors in a long-running cosmic trial, astronomers weighed the evidence as it was presented from time to time. Always hopeful, they nevertheless had to be convinced beyond any reasonable doubt before rendering judgment that stellar parallax had indeed been measured. Henderson was reluctant to present an inadequately prepared case to that court of opinion. If he didn't fully believe the sufficiency of his own data, how could he expect otherwise of his compatriots? He'd rather be second in the parallax race than be the object of professional scorn. In the end, Thomas Henderson decided against immediate publication of his Alpha Centauri work. He would wait for his assistant, Meadows, to complete the supporting observations. *Then* he'd tell the world.

Friedrich Wilhelm Bessel.
Source: *Sternwarte, Universität Bonn.*

12

The Twice-Built Telescope

I was thinking the day most splendid,
till I saw what the not–day exhibited,
I was thinking this globe enough, till there sprang out
so noiseless around me myriads of other globes.
—*Walt Whitman, "Night on the Prairies," from* Leaves of Grass

The winds and waves are always on the side
of the ablest navigators.
—*Edward Gibbon*

My father's desk was filled with mysterious instruments. As a child, I'd slip into his room, slide open the drawers, and play with the brass compasses and dividers, the beaklike reeds of flexible steel, and the strange curves of amber-colored plastic. I had only a vague idea what each was for, but I knew that when my father had escaped the Nazis in 1939, these were among the few possessions he had carried with him.

My father was an engineer. The mysterious instruments were the tools of his trade. They defined him as a professional and, to an extent he readily admitted, as a person. During the 1950s and 1960s, my father designed rocket nozzles, gimbals, and a variety of spaceship components, including what he mischievously dubbed the "shit mitt," a high-tech porta-potty worn by the Mercury astronauts. Later he designed the mechanisms inside machines that make credit cards. He took his ideas and patiently, proudly transformed them into fine-lined blueprints, which others inflated into reality. I can imagine him, like some Viennese John Henry, going head to head against the best computer drafting program of the day.

My father would have found a kindred spirit in Friedrich Wilhelm Bessel. Both were precise, technical men who valued accuracy and sought perfection. Both employed special instruments and applied them, albeit in very different ways, to the investigation of space.

Minden, Germany, 1799. Signs of revolution were everywhere: political, scientific, industrial, economic. It was a tumultuous time throughout Europe, a time of change, a time of war, a time of opportunity for ambitious, intelligent, hard-working young men. Against this churning backdrop, fifteen-year-old Friedrich Bessel, son of a low-level government functionary, made his decision to leave the city in which he was born and seek a brighter future elsewhere. For as long as he could remember, his family had been teetering on the brink of financial ruin. But now, in the new order, it was possible to create his own fortune and control his own destiny.

Bessel dropped out of school in the eighth grade. Whatever he needed to know, in his opinion, he could learn from books. His father arranged an apprenticeship for him in the bustling trade center of Bremen. Located at the junction of major continental trade routes, the 1,000-year-old city had become a visible symbol of the wealth and power of the middle-class merchant. From the banks of the Weser River, which flowed through the heart of the city, Bessel could have almost smelled the salt air of the North Sea some forty miles downstream. Out there was the wider world, a world he had read about, a world he wanted to explore.

On January 2, 1799, Bessel reported for work on Bremen's Papenstrasse at the import-export firm of Andreas Kulenkamp and Sons. The workday ran from 8 A.M. until 8 P.M., six days a week, and on Sundays until noon. Like his fellow apprentices, Bessel spent the hours writing and copying business correspondence. But at night, he pored through the firm's business ledgers and the elder Kulenkamp's collection of books about finance and overseas trade. The more quantitative the material, the better he liked it, for he was a "numbers" man at heart. By the time he was twenty, Bessel had transformed himself into a mathematician specializing in the financial intricacies of business. His extraordinary skill at commercial accounting and mathematical analysis made him a rising star at the Kulenkamp company.

Meanwhile, Bessel continued to dream of the ship that would someday carry him to faraway ports. He taught himself English and Spanish in preparation for future adventures. His table at home had grown cluttered with books on economics, geography, and foreign customs. But it was a single pair of books, Moore's *Epitome of Practical Navigation* and Bohnenberger's *Introduction to Geographical Position-Finding,* that would change his life.

Celestial navigation had evolved from a sailor's art into a highly mathematical science of arcs, angles, sextants, and chronometers that was beyond the comprehension of most Hanseatic sea captains. But not Friedrich Bessel. At first, he wished only to free the Hanseatic trading vessels from their coast-hugging courses and allow them to sail at will across the expanse of the seas. But the more he read about celestial navigation, the more he wanted to know about its astronomical underpinnings. What factors govern the frequency of the tides? Why do planets trace out loops in the sky? How do astronomers determine precise positions of stars? Are all stars fundamentally like the Sun? And so, Bessel set out on an unexpected journey, far from the practicalities of the business world. His life split in two. By day, he carried out his duties at the trading house. But at night, in the solitude of his room, he focused on his new passion: astronomy.

The seeds of Bessel's interest in science had already been sown before he left Minden. At age thirteen he realized that his eyesight was remarkably acute. A star in the constellation Lyra, which to his older brother had looked like a single bright speck, had appeared to Bessel distinctly as a pair. Bessel consulted a map of the heavens; even the mapmaker had drawn a solitary star there. (At the time, he suspected his brother's weakened eyes were the result of undue diligence in law school studies.) Bessel continued throughout his life to measure his eyesight against the double star, Epsilon and 5 Lyrae.

A second nudge toward the sciences had come from the tutor Bessel's father had hired to boost his son's flagging school grades. During a lesson, the tutor had mentioned a device called a "Brennglass," which is simply a lens that focuses the Sun's rays sufficiently to ignite a piece of paper. At Bessel's insistence, the next lesson had been spent trying to sand down a piece of window glass to the necessary convex shape. The lens never set the paper on fire, but it did fan the embers of scientific curiosity already flickering inside young Bessel.

In Bremen, Bessel immersed himself in the writings of Copernicus, Kepler, Newton, and Halley. He became caught up in their densely mathematical descriptions of orbital motion, reportedly limiting his sleep to five hours a night so he could pour all the new information into his brain. Soon he had taught himself how to compute the shape, size, and orientation of a celestial body's orbit from careful analysis of the body's night-to-night movement across the sky. Now he was ready for a challenge. In modern terms, Friedrich Bessel was an "athlete" whose sport was *computation,* a numerical high-jumper who had just decided to raise the bar, confident that he would sail over it. The "bar," in this case, was a real astronomical problem: the orbit of Halley's comet. Almost 200 years earlier, in 1607, Halley's comet had rounded Earth after an absence of three-quarters of a century. The English observer Thomas Harriot (who had been falsely implicated in the famous plot to blow up Parliament) had recorded the position of the comet as it crept across the night sky. Bessel obtained a copy of Harriot's previously unpublished measurements and during 1804 laboriously calculated the comet's true path in outer space. The computations extended over 300 folio pages.

The resultant orbit came out narrow and long; it extended well past Uranus, the most distant known planet at the time. The orbit was also "flipped over," such that the comet moved counter to the direction of the planets. Most important, Bessel realized that his orbit was nearly identical to the one derived by Edmond Halley himself a century earlier. He sensed he had accomplished something significant, but how could he be sure? The trader's apprentice had to find an astronomer.

Turn-of-the-century Bremen was a center of commerce, not astronomical science. It had no major university with a staff of celestial researchers. It had no large observatory like those in Greenwich or Paris. "Write to me now and then, when you have spare time, about intellectual things," Bessel wrote in 1801 to his elder brother, a law student in Berlin. "Learned people are not to be found in Bremen. It appears that science has completely died out in this city. There is only one man here that we can take pride in."

Wilhelm Olbers was the leading scientist in Bremen. A practicing physician, Olbers had been "bitten by the astronomy bug" when he was a youth. Now in his mid-forties, he had equipped the upper floor of his Bremen home as a modest observatory, and could be found there almost every clear night. Patients requested evening house calls at their peril, for Olbers would often peer skyward out the window while they described their ailments. But Wilhelm Olbers was no hack, either as a physician or an astronomer. When he was only twenty-one and studying medicine, he had devised a simplified method for calculating a comet's orbit, the same method that Bessel had used. And in 1802, he had discovered the second known asteroid, Pallas. (Later in life, Olbers innocently triggered the European "comet panic" of 1832, when he announced that Biela's comet would cross the Earth's *orbit* on its way around the Sun. Newspapers reported an impending collision with the Earth itself! Olbers reassured the public by pointing out that the Earth would be a safe 50 million miles away from the comet when the crossing occurred.)

One summer Saturday in 1804, as Bessel relates in his autobiography, he was walking through the streets of Bremen when he caught sight of Olbers strolling nearby. Here was the opportunity to have his work acknowledged by a comet expert—if only he could muster the courage to speak to him. Bessel followed Olbers for a while, mentally "computing" a trajectory that might bring the two of them together. Finally, he set off on a parallel route. His heart pounding, he darted down an alleyway and emerged right in front of the startled Olbers. A stammered apology. A rushed account of his comet calculations. A request for Olbers to review the work. Olbers immediately put Bessel at ease. He would be more than happy to check the calculations. In fact, he sounded positively eager to see the new work. Bessel ran home, grabbed the comet papers from the table, and delivered them to Olbers's house.

Imagine Olbers's reaction as he paged through Bessel's work. Here was a younger reflection of himself, a twenty-year-old amateur comet researcher who had tackled a complex problem for the sheer joy of the challenge. Yet Olbers must have seen much more in the precise computations, thorough explanations, and crisp reasoning, all worthy of a doctoral dissertation. He immediately contacted his close friend, the famous mathematician Carl Friedrich Gauss, to tell him about the aspiring astronomer who had materialized in front of him as unexpectedly as a new comet. Olbers, with characteristic modesty, later described Bessel's recruitment as his greatest service to astronomy.

On Sunday afternoon, the day after he had given Olbers his work, Bessel set off on a long hike to calm his nerves. When he returned, he found a letter and a pair of books beside his door. Bessel's mathematical skill and astronomical knowledge, Olbers had written, were truly remarkable for one so young. The comet orbit had been computed correctly, and if Bessel would factor in the additional observational data recorded in the accompanying books, Olbers would see to it that the final results would be published. In November 1804, Friedrich Bessel's first scientific work appeared in press.

For the next two years, Olbers took charge of Bessel's scientific education, plying him with books on astronomy and advanced mathematics, assigning him more comet orbits to compute, and generally trying to persuade him that his future lay not in the world of business, but in the stars. Bessel considered the unfailingly generous Olbers a second father, and the two men kept up a lifelong correspondence that totaled 360 letters. By 1806, Olbers had landed his young protégé a job in nearby Lilienthal as an assistant to yet another dual-career astronomer, chief magistrate Johann Hieronymous Schröter. The annual salary, Olbers reported, was meager, but Bessel would get his hands on some first-rate astronomical equipment. Simultaneously, the Kulenkamp company offered Bessel a permanent position at a salary seven times what Schröter offered. Bessel had to choose between a frugal existence studying the heavens or a life of wealth, comfort, and prestige as a businessman. Ultimately, the lure of the stars proved irresistible. Bessel packed his bags and on March 14, 1806, headed for Lilienthal. Olbers wrote to his friend Gauss, "I am delighted to report that our Bessel is now completely won over to astronomy. . . . Such a genius . . . will not come my way again."

Bessel later characterized his Lilienthal years as "happy and quiet." Quiet, indeed. After the bustle of Bremen and the Kulenkamp office, life in Lilienthal with Schröter and his elderly sister must have seemed positively bucolic. Naturally gregarious, Bessel took up an extensive correspondence with astronomers throughout Germany, both to keep abreast of activity in the field and to lift himself out of his isolation. His off-hours he spent as always—studying. Sleep remained a low priority. "Take care of your health, my dear overeager astronomer!" the ever-paternal Olbers chided in a letter.

Schröter's telescope was colossal for its time, much larger than anything Bessel had seen in Bremen. Fifteen feet long and a foot wide, the big reflector was supported by a rotating wooden carriage as tall as a house. (In fact, the carriage *was* a house of sorts, complete with door, windows, and exterior staircase.) To the modern eye, the entire assembly looks like a collaborative work of Leonardo da Vinci and Fred Flintstone. To a would-be astronomer like Bessel, it must have been awe-inspiring. Here, finally, he

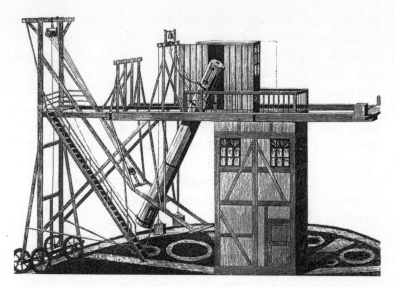

Johann Schröter's reflector telescope, Lilienthal, 1793. From Schweiger-Lerchenfeld (1898).
Source: *Widener Library, Harvard University.*

would have his chance to learn the observer's craft. Over the next few years, Schröter taught him just that. It wasn't long before the pupil's reputation in the astronomical community rivaled that of his teacher.

Bessel's long nights with the telescope probably taught him something else: frustration. To point the instrument, he had to raise or lower its tube by means of a crude rope-and-pulley arrangement, then call to his coworkers to rotate the enormous wheeled carriage. At high magnification, star images in the eyepiece looked blurry, as the effects of every optical imperfection of mirror and lens were likewise amplified. Tracking the movement of celestial objects was difficult, accurately registering star positions nearly impossible. In short, as impressive as it might have appeared, Schröter's telescope was as far from a modern-day telescope as a Model T is from a Mercedes. But in its numerous deficiencies were lessons that left an indelible impression on the young astronomer from Bremen, lessons that would ultimately drive him to reinvent the science of astronomical observation.

Bessel had little interest in mapping the surface of the Moon, as Schröter himself was doing, nor did he wish to endlessly scan the sky on the remote chance of discovering a comet. He wanted to tackle a project that combined precision observing with mathematical analysis. Ironically, Schröter's bulky telescope pointed the way. For centuries, the motions of planets and comets had been gauged with respect to the irregular grid of background stars. Stars had become to the astronomer what modern navigation beacons are to the airplane pilot: fixed landmarks from which to measure the position of a moving object. For example, a subtle shift in the path of a planet might indicate the presence (through its invisible gravitational tug) of an undiscovered planet elsewhere in the solar system. But such a shift cannot be detected if the starry grid against which it is measured is poorly determined. And by the early 1800s, star positions were still not very accurately known.

Bessel envisioned a telescope with unblemished optics, silky-smooth motion, vibration-free mounting, finely engraved coordinate circles. With such an instrument, he could measure the precise positions of stars. Thousands of stars. The task might take decades to complete, but if done right, he could fix the grid of stars to unprecedented accuracy. *Astrometry*—the precision measurement of star positions—was not glamorous work, but to Bessel it must have seemed the most heroic venture he could undertake. It was a labor that would bring him face-to-face with the most vexing astronomical problem of all: measuring the parallax of a star.

Trying to measure a star's parallax with an eighteenth-century-era telescope was like trying to measure a microbe with a ruler. The largest instruments, typified by William Herschel's massive, wood-framed reflectors, resembled cosmic siege engines poised to breach some invisible wall into the heavens. Even the sleek refractors of the day, all polished brass and wood, were insufficiently rigid and too coarsely calibrated to record a parallax. The telescope, that remarkable instrument that had opened up the heavens 200 years earlier, had now become the main factor limiting progress in stellar astronomy. No astronomer appreciated this fact more than Friedrich Bessel.

A telescope, Bessel told an audience in 1840, is an imperfect thing, not to be trusted. Each instrument harbors microscopic defects that reveal themselves only through detailed, systematic observations of the heavens. By modern analogy, a car might look dazzling in the showroom, yet disclose a host of wobbles and squeaks on the test-drive. In Bessel's view, every telescope has to be "built" twice, "once in the workshop of the artisan, from brass and steel, and again by the astronomer, on paper, through the application of necessary corrections obtained in the course of his investigations."

To ferret out an instrument's shortcomings, the astronomer turns inquisitor: Is the telescope's lens at a precise right angle to the light passing through it? Does the lens sag when the telescope is tipped toward a different direction? Are the mount's rotation axles exactly perpendicular? Does the telescope tube warp under the pull of gravity? Is the entire instrument level to the ground and aligned north-to-south? Are vibrations of the astronomer's footsteps transmitted to the instrument? Are the markings on the brass coordinate circles equally spaced? Do the circles themselves contract in the cool night air? In this way, the astronomer-inquisitor tracks down and quantifies every conceivable source of instrumental error, so that their combined effect on a star's measured position can be mathematically nullified. Bessel also described *noninstrumental* factors that conspire to shift the apparent position of a star, the Earth itself being the major culprit. The Earth is an imperfect platform from which to observe the heavens. It hurtles around the Sun, spins, and precesses like a top—and it is enveloped by a curtain of air.

The Earth's atmosphere behaves like an inconstant lens, swelling the pinpoint image of a star. To consistently locate the precise center of this agitated mottle of light poses a serious challenge to the parallax astronomer. The atmosphere also bends starlight that passes through it, progressively tweaking the star's apparent position as the star descends from zenith to horizon. At an angle of 30 degrees above the horizon, the atmosphere crooks a beam of starlight a little over $^4/_{100}$ of a degree before the beam reaches the telescope. Thus, the star appears $^4/_{100}$ of a degree away from its actual position in space, equivalent to the apparent width of a quarter at eighty feet. To the backyard astronomer, such a deviation is unnoticeable. But to a parallax astronomer like Bessel, who was trying to ascertain the star's true position to *one millionth* of a degree, it's practically a country mile. By mathematical manipulation, Bessel effectively had to strip away the Earth's atmosphere, stop our planet from spinning, and halt its revolution about the Sun. Only then would his twice-built telescope indicate the true location of a star.

Bessel continued his work in Lilienthal, while pondering an assault on the parallax problem. As early as 1806, he had made a rash attempt on the distances of several bright stars, but was unsuccessful. All he lacked, it seemed, was a good-enough telescope. Johann Schröter had no need for a costly, high-precision instrument; for lunar mapping, the big reflector was more than adequate. Then during the first week of November 1809, Bessel received a letter, and a universe of opportunity unfolded before him. Friedrich Wilhelm III, King of Prussia, had authorized construction of a new state observatory at the university in Königsberg (present-day Kaliningrad), where Immanuel Kant had once taught. The observatory would be equipped with the finest instruments in the world. And Alexander von Humboldt, the noted explorer, naturalist, and scientific advisor to the King, had recommended Bessel as director of the new facility. The stars, which had seemed so remote only moments earlier, were now within Bessel's grasp.

On March 27, 1810, twenty-six-year-old Friedrich Bessel left Schröter's observatory in Lilienthal to take up his new post in eastern Prussia. He made two stops along the way: first in Minden to bid farewell to his parents and to collect his sister, Amalie, who would accompany him on the trip; and then in Berlin at the studio of Leonhart Posch to have a plaster relief portrait made of himself—a gift for his parents. Bessel arrived in Königsberg on May 11, 1810, bearing with him the plans he had sketched along the way for the new observatory. He had a title and a generous salary, but little else, for Napoleon was waging war throughout Europe, with its attendant economic disruptions. The King had allocated 28,000 Thalers for construction of the new observatory, a substantial sum considering that the university's entire budget that year was 34,000 Thalers. The cornerstone of the facility was not laid until more than a year after Bessel had arrived, on May 24, 1811. Construction could be seen from a great distance; the facility had been sited atop one of Königsberg's highest hills, the Windmühlenberg, named after the windmills that had been razed to make way for the new building. In April 1812, Napoleon's ill-fated Grande Armée marched through Königsberg to engage General Kutusov's forces on the Russian front. The French emperor is reported to have gazed at the half-completed observatory, astonished, and wondered aloud how the king of Prussia could afford such extravagance amidst the turmoil of the times. Meanwhile, back in Lilienthal, Johann Schröter fared less well; the French army confiscated his possessions and burned his observatory and all his records. He died shortly thereafter.

Königsberg Observatory, as it appeared in Bessel's time. From C. L. Littrow's *Kalender für alle Stände* (1848), reproduced in Labitzke (1935).

When completed in 1813, the new Königsberg observatory consisted of a two-story east wing—essentially a house—with living quarters, offices, and a classroom; a slit-roofed west wing where Dollond and Cary transit instruments were mounted; and a midsection with a platform on which portable telescopes could be placed. A domed tower was added in 1829 to house the unusual telescope that would become Friedrich Bessel's eventual "weapon of choice" in his own quest for stellar parallax.

Deskbound during the observatory's construction, Bessel had decided to declare his own vision of "modern" astronomy, founded on his notion of the twice-built telescope. Ironically, this foray into astronomy's future began with a step back in time. Decades earlier, astronomer James Bradley at the Royal Greenwich Observatory had measured the positions of more than 3,000 stars. Bessel obtained a copy of Bradley's meticulous observing journals, then applied a series of mathematical corrections to the "raw" star positions, covering everything from the bending of light in the Earth's atmosphere to instrumental peculiarities Bradley himself had noted. The task consumed Bessel for seven years. In 1818, he published his results as the *Fundamenta Astronomiae pro anno 1755,* a "how-to manual" that guided succeeding generations of astronomers through the complex protocol of "reducing" stellar data. He even tabulated the "personal" errors of contemporary observers, quantifying various idiosyncrasies that affected their measurement of star positions. For example, one of Bessel's observing projects required a precise determination of the instant at which a star traversed a crosshair in the telescope's eyepiece. By analyzing several months of timing data, Bessel concluded that his assistant consistently recorded the star's crossings one second too early, so he added one second to each of his assistant's times. With the *Fundamenta Astronomiae*, Bessel put the astronomical profession on notice: Inaccuracy in observation would no longer be tolerated. Every telescope, every observer, every environment, he declared, leaves its own unique fingerprints on the data it produces. The fingerprints have to be wiped away.

In 1820, Bessel received his first high-precision instrument: a transit circle from the Reichenbach workshop in Munich. The transit circle combined the functions of the mural circle and transit telescope. More stable and accurate than its forebears, the Reichenbach circle permitted the reliable determination of a star's coordinates with a single instrument. The circle was a marvel of precision engineering; even the massive metal trunnions on which the telescope tube pivoted were counterpoised to prevent sagging. With this instrument, Bessel found that he was able to measure star coordinates more accurately than anyone else in the world. Over the next decade, working almost every clear night through the frigid Prussian winters, he accumulated and analyzed multiple measurements of an astonishing 32,000 stars. (Bessel's career nearly ended in 1831, when he was accused of starting a deadly cholera epidemic in Königsberg that claimed over 1,300 lives. Even with his public assurances that the daily rocket signals he launched over the city to mark the time could not cause disease, Bessel's students stood guard outside the observatory to protect their beloved professor from angry citizens.)

It was not until 1834 that Bessel turned his full attention to the challenge of stellar parallax. He knew of many prominent astronomers before his time who had accepted the challenge: Tycho Brahe, Robert Hooke, John Flamsteed, Ole Römer, James Bradley, William Herschel. All had failed. Bessel also knew that the quest for stellar parallax had grown more urgent of late. With the development of truly high-precision telescopes, parallax hunters had come to believe that they finally possessed the tools to succeed. Bessel's contemporaries—Giuseppe Piazzi in Palermo; Giuseppe Calandrelli in Rome; François Arago and Claude-Louis Mathieu in Paris; Baron von Lindenau in Gotha, Germany; and even his former boss in Lilienthal, Johann Schröter—had each claimed victory in what astronomers increasingly perceived as a "parallax race." But instead of glory, the recent parallax competitors gained only the suspicion, if not the contempt, of their colleagues. In a letter to his mentor Olbers on October 30, 1806, Bessel attributes Schröter's "parallax" of the star Alpha Herculis to both faulty instrument and technique.

In one of the more notorious chapters of the stellar parallax story, which Bessel had no doubt followed closely, Reverend John Brinkley at the University of Dublin announced in 1814 that he had measured the parallaxes of Vega, Altair, Arcturus, and Deneb. The results were disputed by England's Astronomer Royal, John Pond, who could detect no annual wobble in any of Brinkley's stars. Since the instruments at Greenwich were superior to those in Dublin, Pond reasoned, Brinkley's parallaxes must be illusory. A decade-long war of words erupted in the pages of the Royal Society's *Philosophical Transactions.* Pond prevailed, dismissing Brinkley with this last parry in 1822: "The history of annual parallax appears to me to be this: in proportion as instruments have been imperfect in their construction, they have misled observers into the belief of the existence of sensible parallax."

Bessel knew that it takes at least a year of dogged observation to detect a single parallax, with no certainty of success. And even if successful in his own eyes, he would still have to convince his fellow astronomers. But having honed his measurement techniques on more stars than anyone else, he was confident that the minute stellar wobble that had eluded his predecessors would eventually reveal itself to him. First, he had to make a crucial decision: Toward which star should he point his telescope? Of the thousands of gleaming specks in the sky above Königsberg, only those nearest the Earth might have a parallax sufficiently large to measure. If he chose the wrong star—a star too far away—he'd waste years trying to detect the undetectable. Here was the same catch-22 that had confronted every parallax hunter before him: To gain the best chance of measuring stellar parallax, a nearby star must be observed; but how can a star be identified as nearby without first measuring its parallax?

Previous parallax astronomers had focused their efforts on the very brightest stars, on double stars, or even on stars that happened to pass conveniently overhead. But as William Herschel had learned, to his disappointment, the stellar species is as varied as the human one. Stars differ widely in their light output, so apparent brightness is an inaccurate gauge of proximity. And many double stars, once assumed to be random line-ups of two stars in space, one near and one far, proved to be true pairs at an indeterminate distance. As early as 1812, Bessel had decided to approach the problem from a different perspective. He sought a swiftly moving star, one that displayed a sizable proper motion across the sky. For if the star showed a proper motion larger than those of others, it was likely to be close to the solar system. Bessel was searching for the northern hemisphere counterpart of Thomas Henderson's fast-moving Alpha Centauri.

Bessel's prayers had already been answered in full by Father Giuseppe Piazzi, a Theatine monk in Palermo. Discoverer of the first known asteroid and himself a member of the burgeoning fraternity of failed parallax hunters, Piazzi was one of the earliest users of the circle-style telescope. His Palermo circle, as it became known, was the brainchild of master craftsman Jesse Ramsden of London. Piazzi had been well aware of Ramsden's reputation for quality. And for delay. Ramsden was easily diverted from ongoing projects to investigate new designs or manufacturing techniques.

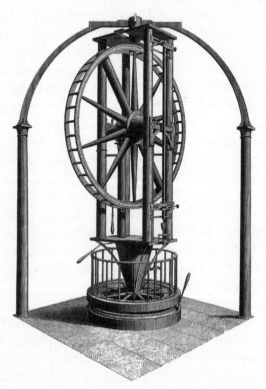

Giuseppe Piazzi's Palermo circle, by Jesse Ramsden. From Pearson (1824).
Source: *Wolbach Library, Harvard University.*

The unfortunate astronomers at Dunsink Observatory in Dublin waited twenty-three years for their Ramsden telescope, which was completed by an assistant after Ramsden died. The King himself fared no better; Ramsden once arrived for a royal function at the preappointed hour—only he was a year late. During an extended stay in London, Piazzi literally plunked himself down in Ramsden's Piccadilly shop to keep the flighty craftsman focused on the Palermo circle. His persistence paid off: Ramsden completed the instrument in 1789, "merely" nine years after Piazzi had ordered it. Unlike Bessel's Reichenbach circle, which was restricted by design to the meridian, the Palermo circle could rotate to access virtually any part of the sky. It was the performance of Piazzi's Palermo circle that persuaded the British government to replace its aging Greenwich quadrants with a mural circle. (And it was German circles, such as Reichenbach's, that subsequently drove the replacement of the Greenwich mural circle in 1851 with the exquisite Airy transit circle, which remained in service until 1954.)

With the Palermo circle, Piazzi assembled his own compilation of stellar positions and motions. Of the more than 7,600 stars in his catalogue, one in particular stood out. In the constellation of Cygnus, trailing the Swan's right wing like a fleck of celestial down, lies a pinpoint glimmer known still by its designation in John Flamsteed's early eighteenth-century catalogue: 61 Cygni. Buried in the rich star clouds of the northern Milky Way, 61 Cygni's orange-tinted glow barely registers on the eye. To the casual sky watcher it is easily passed over in favor of its brighter neighbors: Deneb, the twentieth brightest star in the heavens; and Albireo, a spectacular yellow-blue pair. Yet it was unremarkable-looking 61 Cygni that promised to throw open the door to the depths of space.

Piazzi found that 61 Cygni had moved considerably over the years, positioned fully one-third of a lunar diameter away from where it had been only a century earlier. Its proper motion was 5.2 arcseconds per year, far larger than that of any other star. When Bessel inspected the field of view around 61 Cygni, he saw several fainter stellar specks that had no measurable proper motion whatsoever, convenient fixed markers for an astronomer hoping to discern a telltale annual wobble in 61 Cygni's movement. Indeed, Piazzi's "Flying Star," as it became known, was precisely what Bessel had been looking for. The centuries-old parallax challenge beckoned to Friedrich Bessel. He had the star. He had the expertise. All he needed now was the proper instrument.

Joseph Fraunhofer, 1825.
Source: *Photo Deutsches Museum, München.*

13

Quest for Precision

When a Telescope is pointed at a Star, the least Defect in the . . .
object–glass . . . immediately stares in your Eye, the Stars not
appearing round, but surrounded by false lights, radiating points,
and little flitting luminous accompaniments. . . . If I was an
optician, I think that I would willingly Waltz blindfold and
barefoot among 9 Red–hot Ploughshares laid at unequal
distances from each other, as have all my Telescopes tried
by that truly troublesome test, a Fixed Star.

—*William Kitchiner, early nineteenth-century English
amateur astronomer and telescope collector*

I am easily satisfied with the very best.
—*Winston Churchill*

Shortly after noon on July 21, 1801, while a teenage Friedrich
Bessel labored in Bremen as a business apprentice, gravity
snapped the spine of a house on narrow Thiereckgässchen in
Munich several hundred miles away. With a deafening crash, the
four-story building disappeared in a cloud of dust and was transformed into a pile of rubble. Only days earlier, workmen had tried
to shore up the sagging walls, but the Earth's inexorable tug
brought that effort to folly and even dragged down the neighboring building for good measure.

The house on Thiereckgässchen served as both residence and
workshop to one Philipp Weichselberger, a competent if undistin-
guished maker of mirrors and ornamental glass. Also inside at the
time of the collapse were Frau Weichselberger and a fourteen-year-
old apprentice named Joseph Fraunhofer. The collapse was only the
latest hardship that had descended upon the young Fraunhofer.

Born in 1787 in the Bavarian village of Straubing, Fraunhofer was the eleventh child of a glazier who had managed to eke out a tolerable existence for his family through hard work and disciplined frugality. He was a sunken-eyed, frail-looking boy whose hollow cheeks and aquiline nose were reminiscent of an underfed bird. Formal schooling had been out of the question; it was simply too expensive. Instead he had helped his father in the workshop and otherwise lived an ordinary child's existence. And from the uncomplicated vantage point of youth, his future must have seemed assured: following in his father's footsteps to a secure, if modest, life as a glass cutter. But the trajectory of one's future, young Fraunhofer was shortly to learn, is rarely straight. When he was ten years old, misfortune dealt him a double blow. His mother fell down the cellar steps and died; his father died unexpectedly the following year.

In the weeks after the father's death, the family house and possessions were sold and arrangements were made for the care of the children. Alone and frightened, eleven-year-old Joseph Fraunhofer soon found himself huddled in the back of a mail wagon as it bumped along the road to Munich and to his new future in the Weichselberger household. If Fraunhofer had expected any measure of kindness from his new master, he was sadly mistaken. He was given a dank, windowless room without even a candle by which to read. In fact, Herr Weichselberger, who apparently felt he owned every waking second of Fraunhofer's life, forbade books altogether. Nor did he allow Fraunhofer to attend Munich's Sunday apprentice's school. "Free" time was devoted to seemingly endless chores for the mistress of the house. In short, life with the Weichselbergers was a nightmare that could have spilled from the pen of Dickens. The contract with Herr Weichselberger would last six years. To Fraunhofer, it was a six-year prison sentence.

On the day of the building collapse, just halfway through his apprenticeship, Fraunhofer again huddled alone and frightened, this time in utter blackness beneath the ruins of his master's house. Hearing voices above him, he called out for help. And waited. A crowd gathered to survey the wreckage and thanked God they were not buried under the huge mound of wooden beams and mortar. Even Prince Maximilian Joseph, heir to the Bavarian throne, arrived to personally oversee the gruesome and dangerous search for victims. At his side were the prominent statesman and entrepreneur, Joseph von Utzschneider, and Munich's chief of police. Rescuers quickly pulled Herr Weichselberger alive from the wreckage. Then they heard a muffled cry rising from deep within the tangled mass of debris. They hastened their digging. Four hours later they lifted a heavy beam—and there was Joseph Fraunhofer, the apprentice, dust-covered and shaken, but unhurt. He lay in a small cavity in the rubble, nestled beside a wall of stacked crates, which had prevented the building from descending upon him. To the witnesses, it was nothing short of a miracle. The rescuers pulled Fraunhofer to safety, then resumed their grisly search. Several days later, they found Frau Weichselberger's crushed and lifeless body.

Soon all of Munich was buzzing about Fraunhofer's rescue. The gaunt, dark-eyed apprentice—an orphan, no less—was suddenly a celebrity, a walking, talking symbol of hope in a land too long ravaged by discord and poverty. The statesman, Utzschneider, sought him out; this accidental lightning rod of optimism had to be protected. In conversation, Utzschneider recognized Fraunhofer's keen intelligence and desperate desire to learn. The next time they met he presented the unschooled apprentice with books on mathematics, physics, and optics; he suspected the boy would make better use of them than the lazy students he'd heard about at the university. Even the acquaintanceship with someone as prominent as Utzschneider did not prepare Fraunhofer for the next request for his company: an invitation from Prince Maximilian Joseph himself.

The summer palace at Nymphenburg, two miles west of Munich, covers such an expanse that it's hard to absorb in a single glance. The grounds are magnificent: sprawling lawns, lush gardens, fountains, statues, a reflecting pool. Dressed in his rough clothing, Fraunhofer walked up the wide path toward the five-story, red-roofed palace. This was a world apart from the spartan village where he was born and from the barren room he inhabited now. He climbed the stairs to the palace and was escorted into the presence of the Prince. The two had an amiable chat. Then the Prince gave Fraunhofer eighteen ducats and offered to help him anytime as a father would a son—comforting words to a boy who had lost his own father at age eleven.

Of course, neither prince nor statesman knew what role Joseph Fraunhofer would play in the course of human affairs. Nor did young Friedrich Bessel, as he daydreamed of planets and stars and perhaps even of measuring the cosmos, unaware that this half-starved apprentice from Munich would one day provide him with the instrument to do so.

All across Europe, old production methods were coming down just as surely as the house on Thiereckgässchen. In their place emerged a new manufacturing paradigm based on ingenious machines, scientific analysis, surging market demand, and fierce competition. Late eighteenth-century astronomy was simultaneously root and flower of this revolutionary change. While astronomers' tools benefited from improved production standards, the desire for ever more precise astronomical instruments drove those standards higher. And nowhere in science were the technical demands more stringent than in the determination of star positions. To ensure success in this endeavor, telescopes had to evolve from mere magnifying windows on the heavens into accurate engines of measurement.

The precise determination of star positions was important for more than astronomical reasons. In the early 1800s, stars were the equivalent of the global-positioning-system satellites of today. The celestial grid of the night sky provided the most accurate means of computing terrestrial latitude and longitude, the distance between cities, even the extent of an empire. Thus, astronomy became inextricably linked with geography and cartography—and, by extension, with national affairs. Governments enlisted astronomers in efforts to generate maps that indicated precisely where territorial borders lay. Seafaring nations looked to the stars to help them navigate their ships across treacherous waters. Astronomy—especially positional astronomy—swelled into the big-budget science of the early nineteenth century. Observatories and telescopes became a source of national pride, and the desire to leapfrog a competitor's accomplishments anticipated the Apollo-era rivalry between the United States and the Soviet Union.

The practice of precision astronomy had begun in the mid-1700s with James Bradley at the Royal Observatory in Greenwich. The finest telescope makers of the day had been right down the road in London: George Graham and John Bird, in Bradley's time, and later, Jesse Ramsden and Edward Troughton. By the early 1800s, English quadrants, telescopes, and circles could be found at observatories in Greenwich, Oxford, Cambridge, Edinburgh, Paris, Brussels, Palermo, Moscow, St. Petersburg, Cadiz, Göttingen, Königsberg, Kraków, and Madras. It was a Ramsden instrument that Sir George Everest chose for his 1820s geographic survey of India. J. D. Cassini, director of the Paris Observatory, once characterized English telescope makers as "geometers and scientists" and his own countrymen as mere "workmen." After visiting Jesse Ramsden's workshop in 1788, Cassini wrote: "The fertility of this artist's genius, the perfection of his achievements and his consummate experience in his art, compel me to acknowledge that for now and for a long time it will be difficult to attain or imitate his work." Cassini was mistaken. Within two decades of his prediction, British hegemony in astronomical instrument making faced a severe challenge from the German optical workshops of Reichenbach and Repsold. In Germany, a remarkable confluence of talent, entrepreneurship, and determination served to elevate the art of precision manufacture to levels that far exceeded what had been considered state-of-the-art in Ramsden's day.

Another element in the rise of German optical shops was the country's revitalized university system. Friedrich Wilhelm III, King of Prussia, founded the University of Berlin in 1809 and simultaneously strengthened institutions in Munich, Heidelberg, Bonn, Göttingen, and Königsberg. To an extent never before achieved, German professors involved their students directly in scientific research, thereby nurturing the next generation of scientists. In these early nineteenth-century German educational institutions, we have the model of the modern-day research university.

Critical to the success of these scientific ventures was the alliance between German research scientists and instrument makers. In astronomy, as in most of the physical sciences, theories had grown more complex and could be proved—or disproved—only through the patient accumulation and analysis of quantitative data. Astronomers specified the instrumental precision they required, and craftsmen set about to achieve it. In the parallax quest, astronomers had begun to accept the hard reality of their situation: The long-sought-after stellar wobble was so subtle that its detection required instruments of a whole new level of optical and mechanical sophistication—instruments that no one had yet figured out how to make. In fact, the situation demanded an entirely new type of artisan, one capable of applying scientific research methods to the design and fabrication of instruments: the scientist-craftsman. Cast into this pivotal role was a most unlikely individual. Had you seen scrawny Joseph Fraunhofer walking along the roadside from the palace at Nymphenburg, clutching his princely eighteen ducats, you probably would not have given him a second glance. Yet Fraunhofer would grow up to become the greatest telescope maker the world had ever seen.

To young Joseph Fraunhofer, the collapse of the house on Thiereckgässchen was more than a turning point in his life; it was a rebirth. His unlikely survival had brought about an equally unlikely chain of events that promised to lift him out of his bleak existence as surely as the rescuers' arms had lifted him from the rubble. Although his oppressive apprenticeship to Philipp Weichselberger continued in the new workshop on Kaufingergasse, Fraunhofer's flinty master did permit him—grudgingly—to attend Munich's free school for apprentices and journeymen every Sunday. Of course, Weichselberger had little choice; his apprentice's new "father," he was informed, was the Prince himself.

Fraunhofer's self-education continued. Whenever the master wasn't around, Fraunhofer would open one of his growing collection of books on mathematics and theoretical physics. Sometimes on late Sunday afternoons, he'd head out to the Karlstor woods and read in the solitude there. Fraunhofer also befriended Joseph Niggl, a respected Munich optician nine years his senior who agreed to teach him the craft of grinding glass into lenses. Niggl was as close to a perfectionist as his limited equipment and techniques would allow. He disdained hack eyeglass makers whose idea of grinding a lens was to cement a piece of glass to a wooden dowel and rub it in a bowl of grit. In Fraunhofer, he found a kindred spirit who was as fascinated with optics as he was—although he couldn't make sense of the mathematical gibberish the boy spouted.

For almost three years Fraunhofer, like Friedrich Bessel, struggled to keep up a life split between work and study. The apprenticeship was a straitjacket that Weichselberger tightened at every opportunity. Finally in 1804, Fraunhofer took the money the Prince had given him and bought out the remainder of his contract. Freed now, he dove into his studies, learning about all the new discoveries in physics since Newton and making himself knowledgeable in chemistry, hydraulics, machine design, and other branches of science or engineering he might someday need. He also frequented Niggl's shop to perfect his lensmaking skills.

Yet aspirations don't fill one's stomach. Times were hard, even for those with paying jobs. To support himself, Fraunhofer started a "desktop" business engraving copper visiting cards. His timing couldn't have been worse. Once again war was on the horizon and Munich was occupied by French forces; few of the residents were in the mood to socialize. After six months the visiting card business had gone nowhere. Fraunhofer was too proud to ask the Prince for another handout, but too hungry to survive without a regular wage. Despairing, he trudged back to the door of his former master. Weichselberger, no doubt gratified that the uppity boy had received his just desserts, took him in as a low-paid journeyman. Fraunhofer resumed the mind-numbing routine in the workshop. Like the mortar and beams that had rained down on him three years earlier, the weight of his failure settled heavily upon him. And as before, all he could do was wait for someone to rescue him. This time he waited almost two years.

One day in 1806, Joseph von Utzschneider, the statesman who had befriended Fraunhofer after the building collapse, arrived at the Weichselberger workshop. Fraunhofer could not have been more surprised if Saint Nicholas himself had appeared on the doorstep. He'd assumed that Utzschneider had forgotten him by now. After all, how long could such a prominent citizen be expected to bother with the welfare of a poor glass cutter? But Utzschneider had not forgotten Fraunhofer. He was here, he explained, not to help Fraunhofer but in the hope that Fraunhofer could help *him*. For the past two years, Utzschneider had been on a mission. He'd been seeking a singular person, someone with very specific interests and qualifications, for a key position in his new company. He'd already scoured Europe's universities and optical workshops without success when it occurred to him that the perfect candidate might be living right in his own backyard.

Kings, politicians, and generals throughout Europe wanted detailed, accurate maps. Maps to plan military campaigns. Maps to negotiate treaties. Maps to delineate national borders. The maps of a generation earlier were simply too crude for the power-brokers shaping nineteenth-century Europe. Yet existing maps could not be redrawn until the lands were surveyed. And the lands could not be surveyed without precision equipment: surveyor's transits and small telescopes, transits to measure the relative positions of landmarks, telescopes to pinpoint latitude and longitude.

Utzschneider recounted how he and two acquaintances— Georg Friedrich von Reichenbach, a former military engineer, and Joseph Liebherr, a watchmaker and machine designer—had formed the Mathematical-Mechanical Institute of Munich, a private company that sought to break England's near-monopoly on the production of scientific measuring instruments. To conquer the market and ensure that the institute would sail untroubled through Europe's fickle political waters, Utzschneider and his colleagues had embarked on a two-pronged offensive. First, they produced everything they needed *in-house*: every bolt, every fitting, every lens, even the specialized optical glass from which lenses are made. Second, they recruited a "dream team" of Europe's brightest minds and surest hands, brought them to Munich, and set them to making instruments of unparalleled quality.

But their plan had hit a snag.

The institute's success, Utzschneider likely told Fraunhofer, depended on the fusion of two engineering disciplines: mechanical and optical. The mechanical side of the institute was in good shape; Reichenbach had spent three years in England studying the methods of master craftsmen like Jesse Ramsden, and watchmaker Liebherr had ample experience working with intricate metal parts. The optical side of the business was more problematic. The institute had already hired a well-known Swiss glassmaker to forge disks of optical glass and a Munich optician—Fraunhofer's friend Niggl—to shape these into telescope lenses. But neither man, Utzschneider explained, had more than a tenuous grasp of theoretical optics; they worked largely by old-fashioned, seat-of-the-pants instinct, as artisans had for ages.

A distortion-free optical system consists of a series of compound lenses of various shapes, thicknesses, and compositions. The old trial-and-error fabrication methods didn't guarantee the best results. The institute needed someone who could design complete optical systems *mathematically,* then specify every aspect of their production, from the recipe for making the glass to the configuration of the multiple lenses. The ideal candidate would be well versed in theoretical optics, chemistry, and physics; facile with mathematics; and have lens-making experience. He would become the unifying force between the institute's optical and mechanical disciplines.

Fraunhofer leaped at the opportunity. An interview was quickly arranged with Utzschneider's partners, Reichenbach and Liebherr. Fraunhofer arrived for the interview looking like a nineteenth-century version of the proverbial computer geek, who revels in technical details but can't manage a pair of matching socks. Reichenbach, the spit-and-polish former artillery officer, was unimpressed. The slump-shouldered teenager looked positively disheveled and spoke so softly that he could barely be heard. Was *this* the man who was supposed to spur the institute to worldwide renown?

A local mathematics professor and several of the institute's technicians were also on hand to assess the credentials of Utzschneider's supposed *wunderkind.* They fired question after question at Fraunhofer, who answered them with poise and quiet authority. He'd gone through every line of every book

Utzschneider had given him plus many others he'd bought or borrowed. He'd committed to memory practically everything that was known about optical science. All the years of self-study had prepared him for this moment. By the end of the interview, everyone in the room was impressed. Reichenbach himself voiced what the others must have been thinking: "This is the man we have been looking for."

May 20, 1806, was a joyous day for nineteen-year-old Joseph Fraunhofer. He had seen the last of Herr Weichselberger and was due to start his new job at the Mathematical-Mechanical Institute. Fraunhofer was placed in charge of coordinating the fabrication of optical components: to make sure enough glass was made so every surveying instrument and telescope could be fitted with the proper configuration of lenses. With unbridled curiosity, he swept through the institute, learning every facet of the business. He could be surprisingly persuasive—in fact, relentless—when it came to advocating higher production standards. Even Utzschneider found him overbearing at times.

Fraunhofer improved almost every aspect of the lens-making process and developed a host of scientifically based testing methods. He pushed the technicians to adopt more modern working procedures. Quality and productivity improved dramatically, and with them, Fraunhofer's stature at the institute. Occasionally Fraunhofer joined his friend Niggl at the optical workbench to grind his own lenses. The lenses performed below his expectations. To his astonishment, tests revealed subtle flaws in the optical glass itself. The disks that came from the institute's foundry were nonuniform; some parts of the disks refracted light differently than other parts. With nonuniform glass, a telescope would always show defective images, regardless of how carefully the lens was shaped and polished. Fraunhofer asked to inspect the foundry. This, he quickly learned, was a touchy subject.

Two types of glass emerged from the institute's foundry. Common crown glass (soda-lime glass), the kind used in windows and spectacles, was formed by melting together sand, soda, and lime. Flint glass, also called lead crystal and found in decorative glassware and chandeliers, is produced by adding lead oxide to the mix. The name "flint glass" stems from the early use by English inventor George Ravenscroft of crushed flint in place of sand; flint is no longer part of the recipe. Crown glass and flint glass each refract and disperse light in their own unique way, such that when fitted together they make a superior telescope lens.

Throughout the 1700s, England held a virtual monopoly on the manufacture of flint glass, and little was set aside for export. However, a confiscatory excise tax so decimated the English flint-glass industry that by the late 1820s the government was forced to launch a crash program to rediscover the means of production. Into the vacuum created by the English tax stepped glass makers on the European continent. At the time Fraunhofer took up his post at the institute in 1806, the continent's foundries produced mostly small-diameter glass disks for the lucrative eyeglass and crystal trades. The casting of large flint-glass disks had proved especially troublesome: If the furnace was too cool, air bubbles remained in the finished glass; if the furnace was too hot, the heavy lead oxide sank, making the glass nonuniform. Disks wider than four inches were seldom worth grinding into finished telescope lenses.

By 1800, only one man outside of England had succeeded in producing tolerable flint glass in large sizes—a mercurial clock-cabinet and bell maker from Switzerland named Pierre Louis Guinand. For three decades, Guinand had tested various ingredients, temperatures, oven designs, and stirring methods before achieving his well-earned success. Understandably, Guinand kept his procedures secret. (Since 1773, the French Academy of Science had offered a monetary prize to anyone who could produce large disks of homogeneous flint glass; Guinand never applied because he would have had to reveal his methods.)

The Mathematical-Mechanical Institute needed Pierre Louis Guinand, and Utzschneider was prepared to pay dearly to get him. Guinand was well aware that, as far as optical glass for telescopes was concerned, his was the only game in town. He held out for a lucrative ten-year contract, which included free housing at the institute's optical works outside Munich, a sizable share of the institute's profits, and a clause that he would have to instruct no one outside his immediate family about his working methods. A sweetheart deal for Guinand, but potentially troublesome for Utzschneider and his partners; the institute's entire production pipeline hinged on one contractor whose self-interest was not necessarily their own. As if to confirm their worst fears, just a year later, in 1807, Guinand demanded a *tripling* of his salary plus a generous pension should he ever leave. Utzschneider decided to forfeit this battle if, in the end, he might win the war. He agreed to all of Guinand's terms. But he held fast to one condition: Guinand had to train Fraunhofer to make optical glass.

It was painful for Guinand to admit the intense newcomer to his foundry; from the beginning he had considered the place his exclusive domain. He felt threatened, as well he should have. Fraunhofer did little to placate Guinand. He openly criticized the quality of the glass that emerged from Guinand's furnaces. Fraunhofer was always suggesting ways to improve the glass— raise the oven temperature, provide more ventilation, lengthen the cooling time—when to Guinand's eye the disks looked perfectly acceptable. And when Fraunhofer launched into one of his technical discourses on theoretical optics, the full weight of the young man's intellect rolled over the hapless Guinand like a tank.

In 1809 Fraunhofer was promoted to junior partner. He had a substantial salary, a share of the profits, even his own apprentice, an unschooled local boy named Georg Merz (who would grow up to lead the company). Fraunhofer was now in charge of all forty-eight workers in the optical division—including Pierre Guinand. The meteoritic rise of the twenty-two-year-old did not sit well with Guinand, then in his sixties. Finally in 1813, Guinand had had enough. He packed up and moved back to Switzerland. The institute felt not a ripple at his going. He wasn't needed anymore. Fraunhofer was brewing his own glass in the institute's furnaces, which had been rebuilt to his specifications. The new disks were not only better than Guinand's, they were also unsurpassed in the world. Fraunhofer's stamp was on every step of the institute's production process. His optical systems were simply the best available anywhere. Scientists from all over Europe descended on the workshop to coax Fraunhofer into custom-building an instrument for them. They spoke of owning a *Fraunhofer* the way one speaks of owning a *Stradivarius;* the name had become synonymous with quality.

A year later, in 1814, after Reichenbach had left to start his own company and Liebherr had retired, Fraunhofer was elevated to full partner. Utzschneider saw to the financial end of the institute's business, Fraunhofer to the technical end. The institute was effectively Fraunhofer's to steer in whatever direction he wished. He chose to focus the company's resources in an area that promised to allow the supreme expression of his unique capabilities: the improvement of astronomical telescopes.

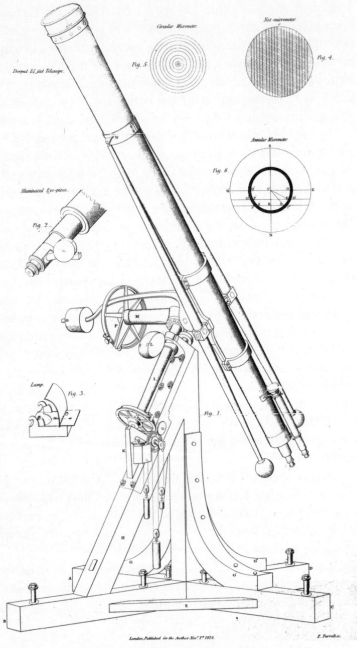

The Dorpat refractor, by Joseph Fraunhofer. From Pearson (1824).
Source: *Wolbach Library, Harvard University.*

14

So Many Grasshoppers

A mere quiver of the telescope tube can make stars leap
away like so many grasshoppers.
—*Rudolph Thiel*, And There Was Light:
The Discovery of the Universe

Many practical men who I have spoken to think that after
Fraunhofer's discoveries, the refractor has entirely superseded
the reflector, and that all attempts to improve the latter
instrument are useless.
—*William Parsons, Third Earl of Rosse*

When I was twelve, I gathered up all my stashed-away coins
and bought my first telescope: a forty-dollar department-store
refractor. My curiosity had been whetted by an astronomy handbook that described in florid detail all the celestial wonders I'd be
able to see with the new instrument. I had already memorized the
constellation patterns, so the layout of the night sky had become
as familiar to me as my own neighborhood. I was eager to discover
the universe that lay beyond the power of my eyes alone.

My first night with the telescope was not what I had expected.
Viewing the night sky from a driveway just ten miles beyond the
lights of Times Square was a challenge; the term "visible to the
naked eye," which characterized many celestial objects in my
astronomy handbook, proved to be a fiction for someone living in
northern New Jersey. Another challenge was the winter chill,
which bit into my fingertips through my gloves. Yet it wasn't the
urban glare or the cold that ruined my evening; these were difficulties I could cope with. No, it was my own *telescope* that utterly
defeated me. When I peered into the eyepiece at my first star, I saw

only a flickering blob of light surrounded by colored fringes. The more I magnified the star, the more irregular it looked. No matter how delicately I tweaked the focusing knob, the star remained blurry. Every time the wind blew, the image hopped about like a luminous jumping bean. And when I accidentally nudged one of the telescope's spindly tripod legs, the star flew out of the field of view. The instrument I'd purchased might have looked like a telescope from the outside and had all the workings of a telescope on the inside, but functionally it was useless. *Caveat emptor!*

Fundamentally, a telescope magnifies distant objects. But that's only half the story. Magnification comes at a price, for every defect in the telescope's optics or mount is likewise enlarged and impresses itself—in varying degrees of ugliness—on the image in the eyepiece. Making a telescope at the dawn of the nineteenth century was a daunting process. Each step was an invitation for yet another flaw to intrude. Metal for mirrors and glass for lenses were brewed in crude furnaces using imprecise methods. The mirrors and lenses in turn had to be ground to precise shapes without the benefit of modern equipment or testing procedures. The long, heavy tube in which they were installed had to be perched on a mounting rigid enough to hold the tube in exquisite balance no matter in what direction it was pointed.

Early nineteenth-century telescopes were generally not as bad as my nightmare of an instrument; still they were not up to the era's most pressing astronomical task: to accurately measure the positions and parallaxes of stars. To unambiguously detect an angular shift as minuscule as a stellar parallax demanded an extraordinary advance in telescopic sophistication. The lessons I learned that frigid night in my driveway about the need for high-quality optics and robust mounts were the same ones that long-ago telescope makers were hearing from their customers. Astronomers wanted better instruments. Yet, for a while at least, no one was building them.

Joseph Fraunhofer set out to remedy the situation. Just as Friedrich Bessel had recognized the need to upgrade the data-reduction process, Fraunhofer recognized the need to upgrade the instruments that gathered the data. Thus, the two men were collaborators in spirit long before they actually met in person. Fraunhofer was uniquely suited to the task. He knew more theoretical optics than perhaps anyone in Europe at the time, had become adept in practical engineering, and displayed an almost obsessive drive to solve any problem he encountered. In his early years at the Mathematical-Mechanical Institute, Fraunhofer had experimented with mirror-based telescopes. He knew about William Herschel's great success with reflectors; hundreds of these instruments were deployed throughout Europe. But he also knew their downside. The solid-metal mirrors deformed whenever the temperature changed, and their reflective surface tarnished under exposure to the elements. Herschel's famous large-aperture reflectors, with their cat's-cradle harness of ropes and pulleys, were too crude for demanding star-position surveys.

Although most refractors in the early 1800s were little more than glorified spotting scopes, Fraunhofer became convinced that this type of telescope held more promise for precision work than did the reflector. Over the next decade, he would elevate refractor performance to such a degree that some astronomers would pronounce the reflector telescope obsolete.

Before the refractor could become the precision measuring engine Fraunhofer had in mind, radical improvements had to be made to its prime component: the main, or *objective*, lens. The objective lens serves to focus incoming light into an image within the telescope tube. An eyepiece magnifies this image to reveal details that would normally escape the limited power of the unaided eye. Ordinary lenses suffer from an inherent defect known as *chromatic aberration*, in which the various colors that comprise a star's light are brought into focus at different distances behind the lens. A telescope's eyepiece can be adjusted to sharpen the image formed by, say, the yellow portion of the starlight, but at the identical setting the "blue" and "red" images will look blurry. As a result, the star can never be brought entirely into focus. The holy grail of telescope making in Fraunhofer's era was the development of an *achromatic* lens, one in which all colors in a star's light are focused to the same point.

Early telescopic pioneers found that chromatic aberration is less noticeable if the telescope's objective lens is made as slim as possible. But such a shallow curvature has an unfortunate side effect: The image is formed very far behind the lens. The slimmer the lens, the farther back the image appears. So although chromatic aberration is moderated, the cure is arguably as bad as the disease. Telescopes more than 100 feet in length began to appear during the mid-1600s. The most outlandish example was the 150-foot-long refractor of Johannes Hevelius, who had to shout instructions to a squad of ropesmen every time he wanted to maneuver the colossal instrument around its ninety-foot-tall mast.

Isaac Newton's response to the chromatic aberration issue was to create a different type of telescope altogether. He replaced the objective lens at the top of the tube with a concave mirror at the bottom. Starlight reflects off the concave mirror to a tiny flat mirror, which directs it through an eyepiece attached to the side of the tube. Thus in 1668 was born the reflector telescope. The development of an achromatic lens effectively ceased after Newton pronounced chromatic aberration the inherent and unavoidable consequence of the refraction of light. Such was Newton's sway among the scientists of his time.

It was not until 1695 that Oxford astronomer David Gregory, inspired by the multicomponent structure of the human eye, suggested that chromatic aberration might be reduced by a compound lens—one consisting of two or more different types of glass carefully fitted together. Yet thirty-eight years passed before anyone followed up on Gregory's idea. In 1733 an English barrister and optical tinkerer named Chester Moor Hall commissioned different opticians to fabricate two small lenses—one a concave flint, the other a convex crown. By coincidence, both opticians subcontracted the work to the same man, George Bass. Realizing that the completed lenses were in fact a matched pair, Bass fitted them together and sighted a faraway object. The all-too-familiar colored fringes were virtually gone from the images. Bass delivered the pair of lenses to Hall, who installed them in a telescope. One peek confirmed for Hall what Bass already knew: David Gregory had been right. A compound lens can cure chromatic aberration.

Bass tried to promote the achromatic lens to established opticians, but found them uninterested. Business was booming, and the opticians were too busy to bother with the newfangled design. The matter lay dormant until 1750 when Bass encountered John Dollond. Former silk weaver, now rising optician, Dollond was firmly in the Newtonian camp when it came to the inevitability of chromatic aberration in lenses. Nevertheless, he was intrigued enough by Bass's report of an achromatic lens that he decided to conduct his own experiments. To his amazement, he found Bass's claim to be true. Further trials led to an optimal configuration: a three-part "sandwich" with a pair of convex crown-glass lenses enclosing a concave flint-glass lens. Word of Dollond's success spread. Astronomers and instrument makers flocked to his workshop. At the urging of his more entrepreneurial-minded son, Dollond claimed the achromatic lens as his own invention; and, to the dismay of his fellow opticians, who had awakened too late to the marketing potential of the new technology, he was awarded a patent and royalty rights. Years of legal wrangling ensued to determine the true inventor of the achromatic lens. Dollond's patent was eventually upheld, as were his royalty rights. "It was not the person who locked his invention in his scrutoire that ought to profit from such invention," Judge Lord Mansfield observed, referring to Chester Moor Hall's early studies of the achromatic lens, "but he who brought it forth for the benefit of the public." Regardless of the legal merits of the case, refractor telescopes had at last shrunk to reasonable lengths without compromising image quality.

Telescopes by Dollond and his associates were highly prized throughout Europe when Fraunhofer sought to surpass them. Dollond refractors were the best available but had remained essentially unchanged for half a century. Lens design, as it was practiced by most opticians, was still largely a trial-and-error affair. Essentially, a variety of lenses were fitted together until a satisfactory outcome was reached. Yet even the few, like Fraunhofer, who approached the problem through mathematical analysis were stymied in their efforts. To design an achromatic lens required detailed knowledge of how glass bends, or refracts, a wide array of individual colors of light. Such data simply did not exist in any usable form. The hitch was the definition of color. Suppose an optician reported the refractive properties of glass for a wide range of colors that included, say, yellow. Specifically what is meant by "yellow"? When I set out to paint our kitchen last year, the man at the hardware store confronted me with a choice of no fewer than twenty-four shades of yellow—with names like Sunshine, Straw, Hawthorne, and Doodlebug. (Doodlebug?) Early nineteenth-century lens designers could not agree on a standard, objective way to specify color. One optician's "Sunshine" might be another's "Straw." Fraunhofer decided to settle the issue once and for all. For inspiration, he looked to the Sun.

The Sun is the most reliable source of light in which virtually all colors are present. Without knowing specifically where it might lead, Fraunhofer began to investigate the Sun's spectrum. It had been almost 150 years since Isaac Newton had first observed the solar spectrum. In his darkened room in Cambridge, Newton had directed a narrow beam of sunlight through a prism and onto a white background. The prism dispersed the light into a blended band of colors (actually, overlapping colored images of the pinhole through which the sunlight entered the room). Newton was able to reconstitute white light from the projected spectrum by passing the colors through a converging lens. He reasoned correctly that white light is an amalgam of the familiar hues of the rainbow.

In 1802 English physicist William Hyde Wollaston refined Newton's experiment, substituting a narrow slit for the pinhole. With this modification, Wollaston saw several fine, dark, parallel lines cutting across the Sun's spectrum. He proposed—incorrectly, it turned out—that the dark lines were natural boundaries separating colors. Fraunhofer repeated Wollaston's experiment with a shaft of sunlight from his window shutters. Instead of viewing the spectrum with his naked eye, as Wollaston had, he snatched a small surveyor's telescope off the shelf and mounted it behind the prism. Fraunhofer had created the forerunner of the modern spectroscope, an instrument that has since become a fixture in observatories and physics laboratories. Hundreds of dark lines were now revealed in the magnified solar spectrum, some narrow and gray, others broad and black. Fraunhofer made a meticulous drawing of the Sun's spectrum, showing a total of 574 lines. He labeled the most prominent ones with uppercase letters, the now-famous *Fraunhofer lines.*

With an improved spectroscope, Fraunhofer saw the identical line pattern in the spectra of the Moon and the planets, indicating that these bodies shine by reflected sunlight. The mysterious lines were also evident in the spectrum of the bright star Sirius, although with different relative darknesses and in a different pattern. Fraunhofer concluded that the lines originated in the Sun and stars themselves, but he'd let the astronomers speculate as to their cause. He was an optical scientist, and although further study of spectra might have been fruitful, he had found what he had been looking for. The Sun's spectral lines, he observed, always kept the same order and relative position. Each line was an indisputable marker of the spectral color adjacent to it. When testing the refractive properties of glass to a particular color of light, lens makers could now specify the color by its associated Fraunhofer line.

Fraunhofer also noted the curious overlap of his solar "D" line with the bright spectral line created by a sodium flame. It would still be several decades before chemist Robert Bunsen and physicist Gustav Kirchhoff, developers of the first fully featured spectroscope, established the connection between Fraunhofer's lines and their counterparts in the laboratory. Each spectral line, they declared, is associated with a particular chemical element. Fraunhofer lines originate when chemical elements near the Sun's surface absorb specific colors of sunlight before the light radiates into outer space. Each element gives rise to a unique spectral-line pattern. For example, the presence of the D line in the solar spectrum means that atoms of sodium exist in the Sun. Thus, the chemical makeup of any luminous celestial object can be determined by identifying line patterns in its spectrum. The French philosopher Auguste Comte was incorrect when he boldly asserted in 1844 that astronomers would never comprehend the chemical or physical nature of the stars. With his proto-spectroscope, Fraunhofer had extended the reach of laboratory analysis to the ends of the observable universe. Earthbound astronomers had gained the ability to "sample" a celestial object by interpreting its light. All in the service of making a better lens.

Fraunhofer applied his newfound knowledge of spectral lines toward an exhaustive study of the refractive properties of glass. With this data in hand, he proceeded to design achromatic lenses mathematically. Everything had fallen into place on the production side: The institute was producing the finest optical glass available, Fraunhofer's scientific lens-testing methods were so sensitive he could detect surface irregularities down to $1/10{,}000$ of a millimeter, the lens-grinding equipment had been improved continually under Fraunhofer's tenure. The achromatic lens had been elevated to a state of near perfection. Yet none of Fraunhofer's advances in lens making would have mattered had the optical assembly been placed on an inferior mount. With few exceptions, telescope makers had concentrated their efforts on a telescope's optics, not on its mechanical components.

Most telescopes prior to Fraunhofer's were mounted like a cannon. The tube could be swung in two directions: perpendicular to the horizon (altitude) or parallel to the horizon (azimuth). These *altazimuth* mounts were relatively simple and cheap to build and gave the telescope access to every part of the sky. However, they were not optimal for tracking celestial objects. As the Earth spins, stars migrate out of the telescope's field of view. With an altazimuth instrument, a star under study must be manually recentered by rotating both axes of the mount. Better is the *equatorial* mount, which is tipped so that one of its axes—called the *polar* axis—is parallel to the Earth's axis of rotation. Now as the Earth spins, a star can be kept within view by slowly rotating the telescope about the polar axis alone. (The altazimuth mount has recently seen a resurgence, as modern computers automatically take care of the complex, dual-axis movements to track the stars.)

Fraunhofer's innovative version of the equatorial mount—the so-called "German" form—was simplicity itself. He mounted the telescope crosswise to one end of a metal axle, counterbalanced by an adjustable weight at the opposite end. He placed the telescope-carrying axle inside a cylindrical sleeve that he attached in turn to the polar axle. The telescope was so finely balanced that it could rotate around either axis at the touch of a finger. Fraunhofer added a gravity-driven mechanical drive that turned the telescope at the precise rate to track the stars. Each axis held a brass circle with etched markings so approximate celestial coordinates could be read off directly. Fraunhofer even developed an elastic support frame for the objective lens, ensuring that the glass would retain its overall shape as it expanded and contracted with temperature. And after every optical and mechanical detail had been taken care of, he made his telescopes beautiful to look at.

Fraunhofer's instruments became legendary. His name came to embody refinement and uncompromising quality. A veritable *Who's Who* of astronomers streamed through Munich to consult with the master optician. Even Czar Alexander I stopped by to urge Fraunhofer to make a precision instrument for Russia. No one quibbled about cost or the years it might take for a telescope to be delivered. They trusted that Fraunhofer would bring each instrument as close as possible to perfection.

In 1819 Fraunhofer began work on a refractor telescope $9^1/_2$ inches in diameter—the largest ever attempted—for Wilhelm Struve, the famous double-star astronomer in Dorpat. About the time it was completed in 1824 to worldwide acclaim, Friedrich Bessel in Königsberg submitted an order for a refractor of his own design, optimized for measuring small angles in the sky. To Bessel's surprise, Fraunhofer replied that his would be the last instrument he would make. Fatigue. Chest pain. Hacking cough. Fraunhofer knew the symptoms of consumption and knew that he might not survive to see Bessel's telescope completed. Perhaps, as some speculated, Fraunhofer's illness stemmed from the long-ago building collapse or the noxious fumes of the glass furnace or his debilitating work schedule. It didn't really matter to Fraunhofer. As his life ebbed away, foremost in his mind was to guarantee that his telescope-making innovations survived. From his sickbed, he issued detailed instructions to his technicians so that every instrument in production and those in the planning stages would be finished to his usual standards.

There was no miracle rescue this time for the former apprentice. Joseph Fraunhofer died on June 7, 1826. He was thirty-nine years old. King Ludwig I declared Fraunhofer's death a severe blow, not just for Bavaria, but for all of Europe. Into the cornerstone of his new palace in Munich, Ludwig laid two pieces of flint glass Fraunhofer himself had cast.

If the legacy of an artist—whether painter, sculptor, or in this case, telescope maker—can be measured by his work, let Fraunhofer's be judged by two of his telescopes. In the 1830s, the Dorpat and the Königsberg refractors were the most sophisticated astronomical instruments on Earth. Placed in the hands of two of the world's most skilled astronomers—Wilhelm Struve and Friedrich Bessel—they heated up the parallax race. For the first time in history, astronomers had in their possession instruments capable of measuring stellar distances.

Fraunhofer is buried at Südfriedhof in Munich beside his former colleague Georg Reichenbach, who had died only weeks before him. The inscription on Fraunhofer's gravestone reads: *Approximavit sidera*—"He brought the stars closer."

Wilhelm Struve.
Source: *Tartu Observatory, Estonia.*

15

The Star in the Lyre

*We Struves cannot live happily without unceasing work, since
from earliest youth we have been persuaded that
it is the best and most useful spice of human life.*
—Jacob Struve, *father of Wilhelm Struve*

*The best kind of heroism is to be found in the
relentless practice of one's profession.*
—*Richard Schickel*, Matinee Idylls: Reflections on the Movies

★ ★ ★

A comet is a wondrous thing: a cosmic interloper, whose realm is neither of the stars nor of the Earth. The comet marches majestically across the sky, appearing in a different spot every night, as the Sun's gravity whips it around in space like a stone in a slingshot. My first comet arrived in 1965 when I was fourteen and living in New Jersey. I remember getting up shortly before dawn and climbing the hill to the aptly named Farview Avenue. Farview Avenue lines the crest of a north-south ridge and, in those days, gave an unobstructed view of the Manhattan skyline to the east. Comet Ikeya-Seki, named for its Japanese codiscoverers, appeared that morning as a nearly vertical chimney of luminous haze, slightly brighter at the bottom, and rising from the spires of the city to a point almost halfway up the sky. The comet just hung there, silent and huge, a simple fact of Nature as powerful in its own way as a mountain range or a hurricane. Now more than thirty years later, I still remember standing alone on Farview Avenue, hands in my pocket, shivering against the chill that ran through my body. The chill arose, I believe, not from the cool morning air, but from the sheer magnificence of the comet itself.

More than a century and a half earlier, the Great Comet of 1807, as it came to be called, materialized in the evening twilight of September. By all accounts, it was a spectacular sight. Week after week, the comet's hazy form lingered in the sky among the planets Venus, Mars, and Saturn. By October, its collective brightness rivaled that of the brightest stars in the heavens and its tail stretched some 10 degrees in length. Among the many observers of the Great Comet were three young men, unknown to one another, whose paths would ultimately cross in the race to detect stellar parallax: Friedrich Bessel, at Schröter's observatory in Lilienthal; Joseph Fraunhofer, at the glassworks outside Munich; and fourteen-year-old Wilhelm Struve, in the Danish city of Altona, near the German border.

Struve's father, a mathematician and rector of the Christianeum Academy where Wilhelm attended school, had hoped that his son would pursue a career in the humanities, perhaps as a teacher or a classics scholar. But Wilhelm had always gravitated toward the sciences, and the allure only intensified with the Great Comet hovering above. The ethereal visitor from deep space had left stars in young Wilhelm Struve's eyes.

Bessel, Fraunhofer, and Struve. Strangers to one another, all three gazed in unison at the Great Comet of 1807. All three shared a curious congruence beyond their common interest in science: The course of each man's life was altered at age fourteen by a circumstance that ultimately, if circuitously, led to a career in astronomy. For Friedrich Bessel, it was the trading-company apprenticeship that brought him to the city of his future mentor, Wilhelm Olbers. For Joseph Fraunhofer, it was the building collapse that lifted him out of anonymity and into the hands of his benefactors. In Wilhelm Struve's case, it was a kidnapping.

Struve was walking through the St. Pauli suburbs of Hamburg, not far from his home in Altona, when he was observed by recruiting agents for Napoleon's Grand Armée. To their eyes, here was a prime conscript: an alert, sturdy-looking, teenage boy, walking alone in French-occupied Germany. They overpowered Struve, dragged him to their safe house, and locked him up in a second-story room. When they had gone, Struve pried open the window and jumped down to the street. He landed unhurt, perhaps in part because of his father's strictly enforced daily regimen of calisthenics, then took off through the streets and alleyways of Hamburg toward his home in Altona.

Struve must have been desperate. At any moment, his kidnappers might discover the empty room and the thrown-back
window. They might even lean over the window ledge, expecting
to see him crumpled in pain on the street below—the two-story
jump should have broken a foot or an ankle. But then surely they
would come after him. He had to cross back over the border into
neutral Denmark before the recruiting agents caught up with
him. He was officially a deserter from the French army and, if cap-
tured in Germany, would be shot. Such was the world in which
Wilhelm Struve was growing up in 1807. "Evil times," his father,
Jacob, had earlier written in his diary. "Evil times."

Wilhelm Struve managed to elude his kidnappers that day. He
made it back safely to Altona, then reported to his father what had
happened. The elder Struve was an astute observer of the political
scene, and the kidnapping confirmed his fears. The situation in
Altona had grown too dangerous for his son. Danish citizenship
no longer seemed to deter Napoleon's recruiting gangs. Relations
between Denmark and England were also deteriorating; he'd
heard talk of an imminent attack on Copenhagen. There was only
one thing Jacob Struve could do: send his son away from the
madness. Wilhelm made preparations to journey 800 miles east to
the city of Dorpat, in the present-day Baltic Sea republic of
Estonia, where his eldest brother, Karl, taught classics at the university. Russia maintained a fragile truce with France, and even if
France were to launch an offensive, Dorpat lay far north of
Napoleon's probable line of attack. For the foreseeable future—
almost an oxymoron in those tumultuous times—Wilhelm
would be safe in Dorpat. And he would feel at home as well, since
Dorpat, like Altona, had a strong German presence. By July 1808,
with a Franco-Prussian treaty in place and further hostilities
unlikely for a time, Wilhelm embarked on the long trip through
Prussia to Dorpat. Among his few possessions were a Danish pass-
port and a letter of reference from a local professor. Like the previous year's Great Comet, now rushing headlong toward the outer
solar system, Wilhelm Struve set foot on his own journey to distant parts.

Struve arrived in Dorpat a month later, moved in with his
brother Karl's family, and enrolled as a student at the university.
Acceding to his father's wishes, he registered for courses in philosophy and philology. On his own accord, he signed up for an
additional course: astronomy. The philosophy and philology

lectures proved uninspiring; after a while, he stopped attending altogether. Instead he devoted his free time to the astronomy course and to a rigorous self-study program in mathematics and science.

In 1809, to lift the burden on his brother's limited resources, Struve became household tutor to the four children of the wealthy von Berg family. (In nineteenth-century Russia, having a foreign tutor for one's children was a status symbol. Even a six-teen-year-old university student like Struve was sufficient. And cheaper.) Struve moved to the sprawling von Berg estate about fifty miles from Dorpat, returning only to take exams and other-wise maintain his student status. For the next two years, he led a glorious life. The von Berg family treated him as one of their own. It was at the many dinner gatherings, hunts, and balls where Struve acquired the poise and social acumen that would serve him later.

Struve received his degree in philology in 1811 and, at his father's urging, considered a secure job as a history teacher at a local school. Only now he felt that same inexorable tug that had drawn Friedrich Bessel away from the world of commerce toward a career in science. Struve disobeyed his father's wishes and returned to the university to take up formal study of astronomy. Dorpat University's lone astronomer, Johann Huth, was chronically sick. Although Huth was a capable scholar, his debilitated state left him barely able to manage his teaching responsibilities, much less run an observatory. Once again, Struve dove into his astronomy texts—including some of the same titles read by Bessel—and taught himself what he needed to know. (Decades later, Struve purchased the extraordinary book collection of Bessel's mentor, Olbers, on behalf of the Pulkova Observatory. On February 5, 1997, Pulkova's rare book library was fire-bombed, apparently by "businessmen" who had been denied permission to build a hotel on observatory property. About 1,000 books and manuscripts dat-ing back to the sixteenth century were incinerated.)

To hone his instrumental skills, Struve bought a precision sextant by Troughton and, during the summer of 1812, set about to map the countryside around Dorpat. Only the drumbeat of war had sounded again. Napoleon's army was marching eastward toward Russia. The local military was on high alert. It wasn't long before Struve found himself confronted by a Russian cavalry patrol that took one look at the fancy sextant and the detailed maps and arrested Struve as a spy. Once again, Wilhelm Struve had run afoul of the military—first with French recruiters, now with the Russian army. He was hauled some ninety miles to a military court in Pernau, where he pleaded his case before the magistrate. A week later, he was released on the condition that he refrain from surveying the area as long as the French army was on Russian soil; it would not do to have such high-quality maps fall into the wrong hands. (Struve seemed to have a penchant for being in the wrong place at the wrong time. In yet another conflict-related incident, Struve happened to be visiting Paris on the day the Revolution of 1830 broke out, and he had to be escorted to safety by the director of the Paris Observatory.)

From his books, Struve had taught himself the theoretical and mathematical aspects of astronomy. With his sextant, he had taught himself the rigors of terrestrial position measurement. Now all that remained was to learn the art of astronomical observation using a telescope. The observatory on Toome Hill was rarely visited by Dorpat's resident astronomer, the sickly Professor Huth. In fact, its major instrument, a transit telescope by the Dollond firm, still lay in its original packing crate.

Lithograph by Louis Höflinger of Dorpat Observatory. Wilhelm Struve's residence appears in the foreground. **Source:** *Anne Lindhagen/Alan Batten.*

I can well imagine Wilhelm Struve's excitement as he pried open the wooden crate that held the Dollond transit. A year after I joined my university, I opened a similar astronomical "treasure chest." I'd been tipped off by an engineering professor that a telescope—a *sizable* telescope—had been sitting in one of the research labs for years, untouched and unclaimed. I made my way to the lab, where I found a big, black trunk, nearly obscured underneath piles of books and assorted equipment. Clearing away the debris and lifting the lid of that trunk, I felt like Howard Carter cracking open King Tut's tomb. Cradled in gray foam was a bright orange cylinder: a reflector telescope, fourteen inches in diameter, worth about $5,000. The new-found telescope became the focus of a fund-raising campaign, which led to the establishment of our campus observatory.

Likewise, Wilhelm Struve had to get the Dorpat observatory up and running. He unpacked the Dollond transit from its crate. Working carefully by hand, he chiseled out cavities in a pair of granite support pillars to secure the telescope's axles. Then, starting in 1813, he used the instrument to determine the observatory's precise latitude and longitude, which he subsequently put forward as a doctoral thesis. On the morning his thesis defense was scheduled to begin, however, word arrived that Napoleon had been defeated in the Battle of Nations at Leipzig. The professors adjourned to join in the nationwide celebration. The next day, they awarded Struve his Ph.D. in astronomy, and a few weeks later invited him to join the faculty. He was twenty years old.

Struve immediately began an extensive survey of double stars using the Dollond transit. His published work attracted considerable attention, both for the sheer number of stars he observed and for the apparent swiftness with which he accumulated results. He worked tirelessly. As his son later noted, "Both mental and bodily fatigue were unknown concepts to him." Within a few years, the same government that had once accused Struve of being a spy requested that he now resume his survey work and finish his detailed map of the Dorpat region. By the 1820s, Struve's domain as official cartographer extended over wide areas of the Russian empire.

During the summer of 1814, Struve returned to Altona to visit his parents and to marry the daughter of a local merchant. On the way, he stopped at the Königsberg observatory and introduced himself to Friedrich Bessel. It was the beginning of both a lifelong friendship and a good-natured rivalry. During their meeting, Bessel likely told Struve about ongoing observing projects, including one that had long occupied his thoughts: an attempt to measure the parallax of 61 Cygni. The attempt ultimately failed, but it might well have stirred Struve's interest in the parallax issue, for shortly after his meeting with Bessel, Struve undertook his own parallax study of a star in the Little Dipper. His published value, based on measurements with the Dollond transit, was not taken seriously by the astronomical community. Bessel and Struve learned a key lesson from their respective failures: Success in the parallax race hinged on further refinement of the astronomical telescope. In the end, both men found what they needed in Munich, in the workshop of Joseph Fraunhofer.

In 1820, while in Munich to check on the progress of a Reichenbach transit circle he had ordered, Struve introduced himself to Joseph Fraunhofer. He was stunned to learn that Fraunhofer was putting the finishing touches on the world's largest telescope lens, $9^1/_2$ inches in diameter. It was to become the core of a revolutionary refractor that Fraunhofer hoped would bring worldwide acclaim to his company. Among the telescope's attributes: a nearly flawless, achromatic lens; a metal-reinforced tube; a hefty equatorial mounting; and steel axles. It would even have a mechanized drive to compensate for the drift of stars induced by the Earth's rotation; point the telescope at a star, engage the mechanical drive, and observe hands-free, while the instrument follows the star's motion across the sky.

The "Great Refractor," as it would become known, was the very pinnacle of what Fraunhofer believed his talent and experience might permit him to achieve. Struve, by this time a double-star expert, viewed the telescope in a more practical, if equally ambitious, light. With an eyepiece micrometer attached to the instrument, he'd have the world's most advanced system for measuring the spacing, orientation, and movement of double stars and would be able to carry out a complete double-star search of the northern hemisphere sky. Upon his return to Dorpat, Struve pleaded with university authorities to purchase Fraunhofer's Great Refractor. That he made a convincing case is echoed in the university's petition to its Chancellor, Prince Lieven: "The opportunity to acquire this instrument, the possession of which would raise our observatory to one of the first in Europe, perhaps will never return." Prince Lieven consented. Dorpat would become home to the world's largest refractor.

Four years later, on November 10, 1824, a procession of horse-drawn carts arrived in Dorpat bearing twenty-two crates of telescope parts weighing altogether 5,000 pounds: the Great Refractor itself. The entire town turned out in celebration. Every mechanical piece had been wrapped in velvet, the big lens itself secured in a spring-loaded box within a box. The only part broken during the long journey was the leg of the Russian naval lieutenant who had accompanied the shipment. A grateful Struve arranged for the lieutenant to recuperate in his house. Four years later, Struve himself stumbled over the telescope's mount and broke his own leg.

Fraunhofer had personally supervised the disassembly and crating of the telescope. (Instruments were always thoroughly tested in Munich before being dismantled and shipped in pieces.) However, he had forgotten to include assembly instructions. Nevertheless, using a drawing Fraunhofer had sent him earlier, Struve managed to piece together the Great Refractor in only six days. The telescope's tube, made of iron-reinforced fir strips, extended over fourteen feet. It was covered in a mahogany veneer that had been polished to resemble burnished copper. To stiffen the top-heavy tube, a pair of tapered metal rods ran alongside, each with a spherical metal counterweight at the lower end. The refractor's equatorial-style mount was formed from heavy oak timbers. Its steel axles carried engraved brass coordinate circles. An elegant curved shaft bore the cylindrical counterweight for the telescope tube, such that the tube was perfectly balanced no matter where it pointed. A series of weights, chains, and pulleys hung from the mount, part of Fraunhofer's trademark mechanical drive, which automatically turned the telescope to track celestial objects. Finally, aware of Struve's reputation as a double-star astronomer, Fraunhofer had included a magnificent eyepiece micrometer to measure the spacing between stars.

In the predawn hours of November 16, 1824, Wilhelm Struve flung open a nearby window and pointed the telescope to the sky. "I stood before this beautiful instrument," Struve commented in a letter to the *Astronomische Nachrichten,* "not knowing which to admire most, the beauty and elegance of the workmanship in its most minute parts, the propriety of its construction, the ingenious mechanism for moving it, or the incomparable optical power of the telescope and the precision with which [celestial] objects are defined."

In his report to the University Council, Struve wrote: "What a difference is seen there! A mountain peak illuminated on the . . .

Moon, which offers me nothing remarkable in the Troughton [refractor], I recognized, by means of the Great Refractor as consisting of six peaks well separated from each other. One of the most difficult of Herschel's double stars I recognized immediately. . . . [Our telescope] surpasses by far any reflecting telescope in the precision of the images." The public was electrified by a report, widely circulated, that one could read the *Journal de Paris* at a distance of 250 meters with the Dorpat refractor. Czar Alexander I was so impressed with the instrument that he sent both Fraunhofer and Struve diamond rings.

Joseph Fraunhofer's optical-mechanical masterpiece was Wilhelm Struve's engine for success. Struve now had in his possession the best general-purpose telescope in the world, the "Palomar" of the 1820s—an instrument capable of measuring stellar positions to an unprecedented accuracy of a few hundredths of an arcsecond, an instrument capable of detecting the parallax of a star.

By the 1820s, there was a pressing need to identify which stars in the heavens were double. Wilhelm Struve and others had confirmed William Herschel's observation that most double stars were, in fact, gravitationally bound pairs—binary stars—not chance alignments of stars widely separated in space. Such binary systems were crucial to the advancement of knowledge about the stars' intrinsic properties, in particular, stellar mass. An isolated star in space yields no direct clue as to its mass. Binary stars, on the other hand, orbit one another according to mathematical laws formulated by Newton and Kepler; stellar mass can be deduced through careful analysis of the orbital motion of the stars. However, it typically takes decades or even centuries for a star to circumnavigate its partner; one astronomer, studying a binary star for a few years, might record only a tiny increment of the entire orbital path. Thus, the sooner these binary stars are identified, the sooner their orbital motions can be pieced together.

With the Great Refractor, Struve undertook a comprehensive survey of the northern hemisphere sky, searching for double stars. This he carried out more or less continuously, even during the blisteringly cold Russian winters. Over the next two years, Struve observed 122,000 stars, sweeping the Great Refractor across the heavens at an astonishing rate of seven stars per minute. In all, he tallied over 3,000 double and multiple star systems, more than two-thirds of them newly discovered. The results of the survey were published in 1827 as the *Catalogus Novus Stellarum Duplicium et Multiplicium,* the capstone of his many previous contributions to the science of double stars. England's Royal Astronomical Society awarded Struve its Gold Medal. The Czar sent him another diamond ring. Only thirty-four years old, Wilhelm Struve was now internationally known.

Struve did not slacken his pace after the publication of *Catalogus Novus.* For the next ten years, he reobserved each of the double stars in his catalogue to precisely determine their celestial coordinates, brightness, and separation between the component stars. In 1837 he released his *Stellarum Duplicium et Multiplicium Mensurae Micrometricae.* (Among the tabulated star systems were sixty-four triples, three quadruples, and one quintuple.) Bessel characterized the massive volume as "a magnificent work ranking among the greatest performed by astronomical observers in recent times."

Among the more than 3,000 stars listed in Struve's *Mensurae Micrometricae* is one that has been familiar to skygazers for thousands of years: Vega, in the constellation of Lyra the lyre. Known to the Akkadians as the "Life of Heaven," to the Assyrians as the "Judge of Heaven," to the Arabs as the "Swooping Eagle," and to the Romans as the "Harp-star," Vega is the fifth-brightest star in the heavens. Its brilliant, bluish-white sparkle forms one apex of the Summer Triangle (with the stars Deneb and Altair), which dominates the northern hemisphere sky from late summer through mid-autumn. On the night of July 16, 1850, Vega became the first star to be imaged on a photographic plate. The photograph was taken at Harvard through a fifteen-inch refractor made by Joseph Fraunhofer's successors, Merz and Mahler.

Struve's interest in Vega stemmed from its apparent pairing with a faint star only forty-three arcseconds away. The two stars had different proper motions—Vega's was much larger—so Struve concluded that Vega and its apparent "companion" were not a physical pair. The companion, he claimed, was an ordinary star whose apparent faintness (a mere $1/10,000$ of Vega's light) was the result of its vaster distance. Therefore, the faint star would display no measurable parallax during the course of the year. Vega's false partner, Struve realized, could serve as a convenient fixed landmark—a "comparison" star—against which to measure Vega's night-to-night position and, ultimately, its parallax. The two stars were spaced closely enough that they fit simultaneously within the Great Refractor's narrow field of view; thus, Struve would be able to apply Fraunhofer's exquisite eyepiece microme-ter to gauge the stars' varying separation during the course of the year. The Great Refractor's severest limitation, in terms of paral-lax research, was its restricted field of view, a consequence of its optical design. Struve was well aware of 61 Cygni's greater promise as a parallax candidate, but the Great Refractor could not at once encompass both 61 Cygni and even its nearest compari-son star, 460 arcseconds away. Thus, he had to make do with the next best candidate he was aware of at the time: Vega.

Struve piggy-backed the parallax study onto his double-star work and his growing administrative obligations. (The Czar had chosen him to design and run the new state observatory in Pulkova, near St. Petersburg, in addition to his regular duties at Dorpat.) On seventeen nights between November 1835 and December 1836, Struve recorded the spacing between Vega and the faint comparison star. From this data, he computed an approximate parallax for Vega of $\frac{1}{8}$ of an arcsecond. Struve knew that seventeen observations were not sufficient for a definitive result, but it had been the best he could do under the circum-stances. Instead of publishing the Vega parallax on its own, he decided to include it in the introduction to his *Mensurae Micrometricae* as part of a comprehensive review of the whole parallax problem.

On July 25, 1837, Wilhelm Struve sat down and wrote a letter to his long-time friend, Friedrich Bessel, informing him that a copy of the new *Mensurae Micrometricae* double-star catalogue would soon be arriving. For the first time in their frequent corre-spondence, Struve revealed that he had been working on the par-allax of Vega and had even deduced a numerical value, although he admitted that the data are "by no means complete enough." Bessel received Struve's catalogue around August 12 and reviewed the Vega measurements. The parallax was indeed preliminary, as Struve himself had pointed out; another year's scattered observa-tions might be necessary to refine it beyond doubt. And that was precisely what Struve intended to do.

A year. Just enough time, if one were to begin right away and observe virtually every clear night, to determine a star's parallax. On August 18, 1837, less than a week after receiving Struve's par-allax announcement, Friedrich Bessel cleared his observing calen-dar and turned his sights on 61 Cygni.

The Königsberg heliometer. From Schweiger-Lerchenfeld (1898).
Source: *Widener Library, Harvard University.*

16

The Subtle Weave

The important thing, however, is not . . . which parallax was
determined first but which parallax actually dispelled all doubts of
the contemporary astronomers that the long–searched–for effect
had finally been found. . . . I believe it is important to distinguish
the result that appeared convincing to the contemporaries of
Bessel, Struve, and henderson, from what we, with more than
a century of hindsight, can recognize as the first successful
penetration of herschel's "barrier."

—*Otto Struve, American astronomer and
great-grandson of Wilhelm Struve*

Out of the strain of the Doing,
Into the peace of the Done.

—*Julia Louise Woodruff*

★ ★ ★

Like many of Joseph Fraunhofer's creations, the instrument
that Friedrich Bessel cast his eyes upon in Königsberg in March
1829 was almost painfully beautiful: a copper-shaded, mahogany-
veneer tube; burnished knobs, gears, and wheels; and a wooden
equatorial mount that descended to Earth through a complex of
gracefully splayed struts and stout beams. A working piece of
sculpture balanced proudly atop its pedestal. Fraunhofer himself
had not lived to see completion of the instrument, which was five
years in the making, but he had instructed his assistants to follow
Bessel's exacting specifications to the letter. Close inspection of
the telescope's upper end would have revealed it for what it was: a
weirdly mutated species of refractor with a split objective lens and
metal choker driven by a pair of thumbscrews down near the eyepiece. A heliometer.

Developed in 1753 by London's John Dollond, the same man who had marketed the world's first achromatic refractors, the heliometer had a single function: to measure small angles in the sky. Dollond's original instrument had been used to gauge the Sun's apparent diameter; hence, the name *helio*meter, after Helios, the Greek Sun god. Bessel had acquired it for a different purpose: to keep track of the angle between 61 Cygni and a fixed comparison star, an angle that would vary regularly if 61 Cygni exhibited a parallax shift.

The new heliometer looked much like a normal refractor telescope, except that its six-inch-wide main lens had been sliced precisely in half with a diamond cutter. The idea of cleaving such a flawless glass is enough to make even a seasoned astronomer shudder. One slip, and the lens might have cracked. But Fraunhofer and his workers rarely slipped. Astronomer John Herschel, who had visited Fraunhofer's workshop when the heliometer was being made, later wrote: "I well remember to have seen this object-glass at Munich before it was cut, and to have been not a little amazed at the boldness of the maker who would devote a glass, which at the time would have been considered in England almost invaluable, to so hazardous an operation."

Each of the semicircular lens pieces had been mounted in its own metal frame and functioned like side-by-side telescopes, forming a pair of images, each half as bright as that of a similar undivided lens. The lens-halves could be slid laterally alongside one another by turning a thumbscrew. A brass scale had been attached from which the relative offset of the two lens-halves could be read. Here even the precision-obsessed Bessel might have shaken his head in awe of the master craftsmen in Fraunhofer's workshop; the divisions on the brass scale were so close together that a microscope was included so Bessel could distinguish them.

In October 1829, Bessel completed the forty-four-foot wooden tower to house his heliometer. He had designed the tower in consultation with his friend Wilhelm Struve in Dorpat, such that the structure not only raised the telescope above the surrounding treetops, but also prevented the intrusion of even the slightest vibration from elsewhere in the building. The central part of the tower's base was filled with five feet of masonry. Atop this were slabs of sandstone and a layer of timbers. Bolted to the timbers were a series of iron-reinforced beams that rose to the upper reaches of the tower and supported the platform on which the heliometer rested. The tower was capped by a dome that turned on twelve rollers and had a shuttered opening to the heavens.

Bessel must have been ticking off the minutes until sunset to try out his new instrument. When darkness fell at last, he pointed the heliometer toward the stars and peered into the eyepiece. The star images were incredibly sharp, according to his 1831 report. The movement of the telescope's axles was smooth and uniform, and the mounting, rock-steady. As expected, when the two lens pieces were precisely fitted together, the heliometer acted like a conventional telescope; the star images from the lens-halves were merged. When Bessel offset the lens pieces alongside one another, their respective images moved apart; the sky appeared as to a drunkard, every star image doubled. Bessel planned to work in reverse: force the images of two *different* stars to merge by moving the lens-halves. The amount of offset between the lens-halves would give the angle between the stars.

In the race to measure stellar parallax, the Königsberg heliometer gave Bessel a dual advantage over Struve's more powerful Dorpat refractor. First of all, the heliometer's field of view was wider than the Great Refractor's, raising the number of potential comparison stars against which to discern a subtle parallax shift. Second, the heliometer's design effectively canceled out the deleterious effects of atmospheric turbulence on the measurement of star positions. Our unsteady atmosphere sets every telescopic star image into a continual Saint Vitus's dance before the eye. Pinpointing a star's position against a fixed crosshair in the eyepiece, as with Struve's instrument, became an exercise in frustration. While the doubled-up images of the heliometer were no less jittery, all the stars danced in synchrony, and it was relatively easy to bring a given pair into coincidence.

Bessel's strategy for winning the parallax race read something like this: Sight the star 61 Cygni and a much more distant comparison star in the field of view. Offset the heliometer's lens-halves until 61 Cygni's image overlaps that of the other star. Read the amount of offset on the brass scale. Repeat the procedure several times every clear night for at least a year. The comparison star will appear fixed, while 61 Cygni's annual parallactic dance will put a tiny weave into its otherwise beeline motion across the sky. Measure the size of the weave. Compute the distance to 61 Cygni. Last, and just as crucial as any numerical result, confirm that the weave achieves its maximum extent in *June* and again in *December*, as it would for a star in the direction of the constellation Cygnus.

The weave *must* be at maximum in June and December; otherwise, the entire body of work will be suspect.

In fact, 61 Cygni is a double star, not the kind William Herschel and Wilhelm Struve had been seeking—an *apparent* double, with one star near and the other far—but two stars equidistant from Earth, bound to each other by their mutual gravitation. More than twenty years earlier, Bessel had discovered that both members of 61 Cygni had identical proper motions; they move in concert across the sky, a strong indication that they are physically associated: a binary star. The binary nature of 61 Cygni meant that Bessel had two nominally identical parallaxes to measure, each providing a check against the other.

The heliometer was notoriously difficult to use with any degree of consistency, and in hands less skilled than Bessel's, it might have generated questionable results. Bessel spent five years applying the dictates of his "twice-built telescope" concept toward the Fraunhofer heliometer. Five years of testing, calibrating, assessing the instrument's overall reliability. Five years during which he could have gotten a jump on everyone else and attempted a stellar parallax measurement. But Bessel's top priority at the moment was to satisfy his most ardent skeptic: himself. Until he was one hundred percent certain that the heliometer was accurate, he would not join the parallax race. Finally, in September 1834, with the heliometer now a virtual extension of his eyes and brain, Bessel set out to measure the parallax of 61 Cygni. Only a few months into the year-long project, however, he was forced to stop. The comparison star he'd chosen was too faint and would disappear from view every time the weather turned less than ideal. He would have to start all over with a brighter comparison star. But it was impossible to devote the next twelve months to continuous parallax observations. Bessel had already agreed to spend three months in Berlin to research the physics of the pendulum. Also, Halley's comet was due for its once-in-a-lifetime swing past Earth, an event that would claim every clear night at the observatory. And the government had enlisted Bessel to measure the length of a degree of latitude in Prussia.

It was not until August 18, 1837, that Bessel rejoined the parallax race, spurred on, he later wrote, by the unexpected news that Wilhelm Struve, in Dorpat, had attempted Vega's parallax with his own Fraunhofer refractor. Struve's seventeen observations—carried out between November 1835 and December 1836, but with a six-month gap in the middle—were summarized in the introduction to Struve's 1837 double-star catalogue, *Mensurae Micrometricae*. From these measurements, Struve deduced a preliminary parallax of one-eighth (0.125) arcsecond, but admitted that the number was highly uncertain. Perusing the Vega results, Bessel found no conclusive evidence for stellar parallax—the position measurements were simply too few—but he evidently believed that Struve was getting close. On October 18, 1837, Bessel wrote to his old mentor Olbers: "I think Struve has taken the lead, for he has made an attempt which, though not as yet a complete success, nevertheless seems to offer good prospects."

Now Bessel delved into his own parallax work like an astronomer possessed. This time, he chose as fixed landmarks not one, but two comparison stars against which to gauge the movements of 61 Cygni's components. Both comparison stars were relatively bright, a lesson he'd learned from his abortive 1834 parallax attempt. Trying to wring every ounce of precision from the heliometer, he observed 61 Cygni ten, twelve, sometimes sixteen times each night, averaging the measurements for greater accuracy. If Frau Bessel had longed for the nighttime companionship of her husband, her only hope was to pray for clouds to roll in.

The heliometer's dark, circular field of view probably occupied Bessel's thoughts during the daytime as well. At the center, a pair of ruddy, orange pinpoints of light—61 Cygni itself—and, forming a near right angle not far away, the two comparison stars. A quartet of stellar specks whose relative positions seemed frozen from night to night, but which might show a telltale variation over the course of a year. By October 1838, Bessel had accumulated hundreds of reliable position measurements, derived from thousands of individual observations, enough to satisfy even his own high standards. He scrutinized the long list of numbers that tallied the night-to-night position of 61 Cygni, and indeed it was changing in just the way he had hoped. The weave in 61 Cygni's proper motion was subtle, but unmistakable. As the Earth headed toward one side of its orbit, 61 Cygni appeared to stray in the opposite direction. Months later, when the Earth had swung to the other side of its orbit, the star obediently repeated its mirror-image shift. Most important, the weaving motion reached its maximum in June and December, as predicted. Friedrich Bessel knew he had reached the end of his quest. A faint glimmer of light near the wing of the Swan had become the first milepost in the seemingly endless ocean of space.

Bessel's results appeared in the *Astronomische Nachrichten* for December 1838, under the title, "Determination of the distance of the 61st star of the Swan." Simultaneously, he notified John Herschel, president of England's Royal Astronomical Society, of his discovery. Ever attentive to detail, Bessel composed the letter to Herschel in his native German, so as to "secure [his] meaning from indistinctness." He relied on Herschel to translate his carefully chosen words accurately into English. After all, many others before Bessel had claimed to have measured a star's parallax; astronomers had to understand precisely his observations and his mathematical analysis if they were to believe him. With characteristic modesty, Bessel prefaced his letter, "Having succeeded in obtaining a long-looked-for result, and presuming that it will interest so great and zealous an explorer of the heavens as yourself, I take the liberty of making a communication to you thereupon. Should you consider this communication of sufficient importance to lay before other friends of Astronomy, I not only have no objection, but request that you do so." This for news the astronomical community had anticipated for nearly 2,000 years.

Bessel closed his letter to Herschel, "I have here troubled you with many particulars; but I trust it is not necessary to offer any excuses for this.... Had I merely communicated to you the results [of the parallax observations], I could not have expected that you would attribute to it that certainty which, according to my own judgment, it possesses." The extraordinary number and precision of Bessel's observations left no doubt among astronomers that the first stellar parallax had indeed been recorded. Some astronomers might quibble with the "particulars," as Bessel had called them, but the end result was indisputable. John Herschel's sole "particular" was whether Bessel had taken into account the effect of temperature on the calibration of the heliometer screw. Bessel, of course, had. Herschel laid his own opinion before the Royal Astronomical Society: "[T]he results have been placed before you: —*oculis subjetca fidelibus*. If all this does not carry conviction along with it, it seems difficult to say what ought to do so."

The parallax angle of 61 Cygni proved to be exceedingly small: about one-third (0.314) of an arcsecond, equivalent to the apparent size of a Manhattan taxi cab as viewed from Mexico City. Bessel demonstrated mathematically that his result was subject to an experimental uncertainty of less than 5 percent. The star's computed distance came out to about 660,000 astronomical units, that is, 660,000 times farther than the Sun, or more than 60 *trillion* miles. The remoteness of 61 Cygni was not unexpected; astronomers had come to realize that if stars were significantly closer, their parallaxes would not have eluded detection for so long. Nevertheless, the huge number—*real,* this time, not conjectured—stirred the imagination, as scientist and nonscientist alike tried to comprehend in their own minds the sheer vastness of the universe. Far-flung as it was, 61 Cygni's apparent sky movement of 5.2 arcseconds per year translated into an actual velocity through space of almost 170,000 miles an hour. Father Piazzi's "Flying Star" had lived up to its billing.

Bessel basked only briefly in all the attention. He dismantled the heliometer, checked the workings of every piece, then reassembled it and began to observe 61 Cygni all over again. In just over a year's time, he tallied more than 400 additional positions. The weave in the star's motion yielded a parallax angle of 0.348 arcsecond, a near match to the original, considering the difficulties associated with this type of measurement. With this confirmation, Bessel was finally satisfied.

Trailing Bessel's triumphant announcement by only two months, Thomas Henderson published his parallax for Alpha Centauri: just over one arcsecond, which placed the star less than 200,000 astronomical units away, or about one-third as far as 61 Cygni. Henderson had completed his observations while at the Cape in 1833, before Bessel had even started. For reasons that are not entirely clear, Henderson delayed the "dirty work" of data analysis until he returned to Edinburgh to take up his new post as Scotland's Astronomer Royal. Even then, he was in no rush to complete and publish the work. In all likelihood, Henderson was suspicious of his own measurements with the Cape's second-rate mural circle and awaited confirming data from his assistant, Lieutenant Meadows, using a transit instrument. Henderson might have also believed that, with Bessel's success now fresh in their minds, astronomers would be more receptive to a less-than-rigorous Alpha Centauri parallax.

Meanwhile, Wilhelm Struve used the Great Refractor in Dorpat to add ninety-six new measurements to his preliminary work on Vega. By late 1839, he concluded that Vega's parallax was about one-quarter (0.261) arcsecond, with about 10 percent uncertainty, placing the star some 800,000 astronomical units from Earth. Struve's revised parallax was about twice as large as the preliminary value he'd published in 1837, casting doubt on the whole enterprise.

Modern measurements have reinforced Bessel's reputation as a meticulous observer. The true parallax of 61 Cygni is 0.287 arc-second, within 10 percent of Bessel's original value, placing the star about eleven light-years from Earth. Henderson and Struve were much farther off. Alpha Centauri is about 25 percent more distant than Henderson had thought, and Vega almost twice as far as Struve's final estimate.

In retrospect, Struve's assault on Vega was by far the most challenging of the three. Struve's nemesis in this case was geometry: Vega's greater distance means it possesses the smallest, and hardest-to-detect, parallax. He'd been aware from the outset that 61 Cygni, with its large proper motion, held more promise of a measurable parallax than did Vega. But the Great Refractor's field of view was too narrow to encompass 61 Cygni and its widely separated comparison stars. Thus, Struve focused his efforts on Vega, with its virtually adjacent comparison star. The easiest parallax target of all was Alpha Centauri, which we now know is the nearest star beyond the solar system. But Alpha Centauri is visible only from the Earth's southern hemisphere, and could not be seen by either Bessel in Königsberg or Struve in Dorpat.

Although some astronomical revisionists have sought to crown Struve winner of the parallax race, Struve himself never claimed that honor. In an 1848 paper, he acknowledged that Bessel had preceded him in obtaining the first definitive parallax of a star, and characterized Bessel's work on 61 Cygni as "one of the greatest discoveries of our century." Bessel and Struve maintained their good-natured professional rivalry, each confident about his own abilities and delighted about the other's successes. Wilhelm Struve's great-grandson, American astronomer Otto Struve, wrote in 1956, "among the memories transmitted to me by my family, the brightest was the high recognition accorded to Bessel by my great-grandfather. . . . There was never any quarrel between these two astronomers, and they maintained the closest bonds of friendship. . . . It is difficult to find fault with those accounts that attribute to Bessel the first determination of a fully convincing stellar parallax."

More than anyone else, Friedrich Bessel ushered in the modern age of observational astronomy. His advances in celestial observation and data analysis were as fundamental to astronomy as was the introduction of astrophotography in the late nineteenth century or the advent of the computer within recent decades. In fact, it's easy for me as a professional astronomer to picture Friedrich Bessel in a modern observatory, surrounded by all the sophisticated electronic hardware that present-day observers routinely use. He'd simply acquire whatever high-tech knowledge was necessary, then set off on some challenging project. For behind all the fancy equipment, the basic methods of precise astronomical observation are largely the ones that he himself invented.

Over a span of months, a trio of astronomers had vaulted the gulf of space and planted three flags in the previously uncharted territory of the cosmic third dimension. A modest claim, to be sure, on what appeared to be a virtually unlimited universe. Yet it offered to astronomers the same sense of excitement that Earthly explorers feel when they set foot on a long-anticipated shore. That the journey had taken some 2,000 years made the hard-won success all the more sweet. Empowered by the ingenuity and tenacity of three of their own, astronomers expected a bright future in which star after star would yield to their instruments.

In 1841, the Royal Astronomical Society awarded its gold medal to Friedrich Bessel. John Herschel addressed the members: "I congratulate you and myself that we have lived to see the great and hitherto impassable barrier to our excursions into the sidereal universe, that barrier against which we have chafed so long and so vainly . . . almost simultaneously overleaped at three different points. It is the greatest and most glorious triumph which practical astronomy has ever witnessed." Herschel next made a bold prediction: "The barrier has begun to yield, it will speedily be effectually prostrated." As if to underscore that sense of confidence, Thomas Henderson announced a preliminary parallax of a quarter-arcsecond for Sirius, the brightest star in the sky. But after that, the cosmic "barrier" seemed to reimpose itself, and despite massive effort by observers worldwide, no other parallaxes were detected for decades. Yet even this failure delivered new insight. The majority of bright stars above our heads are not solar system neighbors with measurable parallaxes; they are far-flung giant stars that pour forth thousands of times more energy than common stars like our Sun. The stellar species, astronomers learned, was even more diverse than they had imagined.

Wilhelm Struve went on to become director of the Pulkova Observatory, which he turned into the most advanced astronomical facility in the world. In all his frenzy of observatory work and European travel, Struve somehow managed to produce eighteen children in two marriages, out of which grew an astronomical dynasty that spanned four generations. Thomas Henderson left the Cape Observatory in 1833 and the following year became Astronomer Royal of Scotland and Regius Professor at the University of Edinburgh. Over the next decade, in addition to reducing the Cape data, Henderson secured accurate positions of some 60,000 stars in the northern sky.

Friedrich Bessel remained at Königsberg for the rest of his life, having turned down the prestigious directorship of the Berlin Observatory. But in June 1842, Bessel finally embarked for England, fulfilling the dream that he'd had while a young apprentice in Bremen—that of visiting foreign ports. He was feted in Manchester at the Congress of the British Association for the Advancement of Science, after which he set out on an eleven-day tour of England and Scotland. Prominent on Bessel's itinerary was a two-day stay at the home of Association president John Herschel at Collingwood in Kent. He told Herschel of the possible existence of an eighth planet, based on irregularities he had noted in the orbital motion of Uranus, the planet that Herschel's own father had discovered. (Neptune was "swept up" four years later in 1846 by Johann Galle in Berlin.) Bessel then traveled to Edinburgh to meet fellow parallax explorer Thomas Henderson. On July 4, 1842, Henderson presented Bessel a daguerreotype of Edinburgh's Calton Hill Observatory, which he had inscribed on the back and which today resides in the observatory archives. Henderson reportedly described his meeting with Bessel as one of the highlights of his life. Two years later, in 1844, Thomas Henderson died of heart disease at age forty-six. No portrait of him has ever been found.

Bessel would outlive Henderson by only two years—years that were marked by debilitating illness. Yet Bessel did not slacken his pace until close to the very end. Work seemed to energize him. Emerging from a several-month-long bout of headaches and chest pains, Bessel closed a September 1843 letter with these words: "*Nun wieder in die Sternwarte, denn es ist wundervoll heiter!*" Here the German word *heiter* can be translated in two ways. Perhaps Bessel was referring to the weather: "Now again into the observatory, for it is wonderfully clear!" Or perhaps, in view of the ongoing ordeal with his health, he had intended the alternate meaning of the word *heiter:* "Now again into the observatory, for it is wonderfully serene!" Then again, to the astronomer, a cloudless night is serenity itself.

Daguerreotype of Friedrich Wilhelm Bessel, September 15, 1843, by Ludwig Moser. **Source:** *Photo Deutsches Museum, München.*

Before his death in 1846, Bessel made another profound discovery—and a prediction. The proper motion of Sirius, according to Bessel's own observations, is distinctly erratic, as though the star is alternately pushed and pulled by some hidden hand. Sirius, he surmised, has an unseen partner of considerable mass, and the pair twirl about each other like dancers, locked in a gravitational embrace. The "dark" star didn't reveal itself in the telescope, but Bessel was certain of its existence. "We have no reason to suppose that luminosity is a necessary property of cosmic bodies," he wrote to scientist and historian Alexander von Humboldt. "The visibility of countless stars is no argument against the invisibility of countless others." Not until 1862 did American telescope maker Alvan Graham Clark spy Sirius's elusive companion. It was a star of exquisite strangeness, even by modern standards, with a Sun's mass compressed into an Earth's volume. A *white dwarf* star. To an astronomer with Bessel's vision, even the invisible is rendered plain.

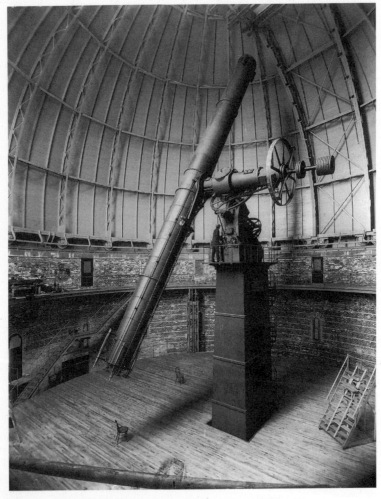

Forty-inch refractor, Yerkes Observatory.
Source: *Yerkes Observatory.*

Epilogue
A Drink from the Well

The cardinal truth emerging from these inquiries is that of the
extreme isolation of the solar system. A skiff in the midst of a vast
unfurrowed ocean is not more utterly alone.

—*Agnes Clerke*, The System of the Stars, *1905*

. . . and thus we die,
Still searching, like poor old astronomers,
Who totter off to bed and go to sleep,
To dream of untriangulated stars.

—*Edwin Arlington Robinson, American poet (1869–1935), "Octaves, XI"*

With the parallaxes of three stars—61 Cygni, Vega, and Alpha
Centauri—having been snared in such swift succession, and with
thousands of tantalizing parallax targets gleaming overhead, the
stellar realm beckoned to a new generation of parallax hunters.
For the most ambitious, the instrument of choice was the
heliometer. Friedrich Bessel's thorough explication of the principles and practice of heliometer measurement paved the way for
the instrument to become the premier stellar-parallax detector of
the nineteenth century. Joseph Fraunhofer's successors, Merz and
Mahler, constructed a near-duplicate of the famous Königsberg
heliometer for the University of Bonn in 1848 and an even larger
heliometer for Wilhelm Struve's staff at the imperial observatory
in Pulkova, Russia. Repsold heliometers were installed at Oxford's
Radcliffe Observatory and at observatories in Stuttgart, Leipzig,
Kapstadt, Göttingen, and Bamberg. Meanwhile, in the southern
hemisphere, David Gill at the Cape of Good Hope began to measure

stellar parallaxes with a seven-inch-wide heliometer. His one-time assistant, W. Lewis Elkin, started the first U.S. heliometer parallax program, at Yale, in 1885. In 1896, the largest heliometer ever made—$8\frac{1}{2}$ inches in aperture and ten feet long—was installed at the Kuffner Observatory in Vienna, where it resides today.

As before, the parallax hunters concentrated their efforts on the two categories of stars most likely to present a measurable parallax weave: the brightest stars and stars of large proper motion ("flying stars," like 61 Cygni). A few of the brightest stars, such as Sirius and Procyon, proved to be relatively nearby. However, the majority—including Canopus, the second brightest star in the heavens; Rigel and Betelgeuse in Orion; and 61 Cygni's constellation-mate, Deneb—displayed no parallax whatsoever. These stars evidently lie a great distance from the Earth; it is only their prodigious luminosity that makes them stand out in the night sky. They are swollen giant and supergiant stars: "Brobdingnagian orbs," as characterized by historian Agnes Clerke; "their magnificence defies the realizing efforts of imagination."

Thus, stars that merely appear bright to the eye did little to improve the parallax harvest during the nineteenth century. On the other hand, stars that move relatively quickly across the sky, regardless of whether they are bright or faint, turned out to be much more productive. Positional astronomers, using ever more precise transit-circle telescopes, uncovered many faint stars that had shifted substantially since James Bradley had catalogued their positions a century earlier (just as Bradley's predecessor, Edmond Halley, had noticed shifts of stars since the ancient days of Hipparchus). Nearly every one of the large-proper-motion stars proved to have a detectable parallax. Astronomers thus confirmed their long-held suspicion that these "flying stars" fly through space no faster than the average star; they *appear* to move more quickly only because they are closer to the solar system. (Later, others were found that do, in fact, move quickly.) The twin characteristics of faintness and proximity paint a picture of a stellar population quite unlike the luminous giants and supergiants: Most of the Sun's neighbors are relatively small and emit but a fraction of the Sun's energy. At the dawn of the nineteenth century, William Herschel's binary-star observations had pointed to the diversity of the stellar species; now, as that century was drawing to a close, Herschel's conclusion was amply confirmed by the parallax hunters.

The end of the nineteenth century also brought disappointment. The parallaxes of fewer than 100 stars had been obtained, with different observatories often getting widely divergent results for the same star. Disputes simmered as to who was right and who was wrong. Every telescopic measurement carries with it a degree of uncertainty, parallax measurements more than most. Scientists do their best to minimize the impact of these so-called random errors on the final result; they repeat observations many times, then average the data. If the errors are truly random, sometimes the measurement will come in too big and sometimes too small, but the average of many such measurements should be close to the true value being sought. What affected the nineteenth-century parallax effort was more insidious: "systematic errors," a hidden skewing of instrument or observer that causes measurements to always turn out either too large or too small. For example, a mal-adjusted bathroom scale might gauge your weight with tenth-of-a-pound precision, yet the displayed weight will be wrong; the scale is precise, but it is also skewed. Similarly, an observatory might determine a star's parallax with great precision, yet the result might well be fallacious.

Parallax astronomers adhered as best they could to Friedrich Bessel's principle of the "twice-built telescope," but they ultimately failed to account for all of their instrumental and personal peculiarities. Sitting hour after hour in the solitary stillness of the observatory, where the only sounds are the steady whir of the telescope drive and the occasional creak of the observer's chair, where one's fingers grow stiff in the frigid night air, where fatigue weighs both on eyelids and mind, and where the starry image in the eyepiece quavers incessantly, it is easy to understand how astronomers might be subject to, in historian Agnes Clerke's phrase, "certain idiosyncrasies of perception."

By the end of the nineteenth century, a gloom had descended on the entire parallax effort. From the moment Bessel had announced the first stellar parallax in 1838, astronomers had envisioned a more or less steady march toward the ultimate goal: mapping the complete stellar distribution around the solar system. Now, after six decades of intensive observation, that goal must have seemed as remote as the stars they were trying to measure. Some researchers felt that the era of parallax measurement was effectively over and that astronomers had been defeated by the sheer immensity of the realm they were attempting to chart. But such pessimism proved to be unfounded. The number of stellar parallaxes was about to explode, from barely 100 to almost 2,000 just twenty-five years later. Like their predecessors earlier in the century, a new generation of parallax hunters hitched a ride on a wave of technology. Gone was the heliometer. In its place: the camera.

On January 7, 1839, just two months after Friedrich Bessel's heliometer observations of 61 Cygni appeared in print, French astronomer-physicist François Arago announced that stage designer Louis Jacques Daguerre had discovered the means to record the visible world on specially treated copper plates. The announcement created a sensation, and before long, others were exposing their own "daguerreotypes." Bessel himself sat for a portrait in 1843, perhaps unaware that the photographic process spawned by the daguerreotype would someday render his heliometer obsolete.

Observers were stunned by the daguerreotype's realism. "We have seen the views taken in Paris by the 'Daguerreotype,'" announced *The Knickerbocker* magazine in December 1839, "and have no hesitation in avowing, that they are the most remarkable objects of curiosity and admiration, in the arts, that we ever beheld. Their exquisite perfection almost transcends the bounds of sober belief." The public was equally awestruck, according to one early daguerreotypist: "People were afraid at first to look for any length of time at the pictures he produced. They were embarrassed by the clarity of these figures and believed that the little, tiny faces of the people in the pictures could see out at them, so amazing did the unaccustomed detail and the unaccustomed truth to nature of the first daguerreotypes appear to everyone."

It did not take long for astronomers to recognize the research potential of the new technology. Photography would be more efficient than visual observation, for each photographic plate would record the images of many stars simultaneously. It would also be more convenient: photographs could be taken at night, then studied by any number of researchers during the daytime. Through time exposure, photography promised to reveal celestial objects too faint to be seen through the telescope by eye; parallax astronomers could envision hosts of newly discovered comparison stars against which to measure a target star's position. And following initial inspection, photographic plates could be archived to form a long-term record of the night sky. (Most major observatories maintain such archives. Harvard Observatory's "plate stacks" house almost 500,000 celestial photographs dating back more than 100 years.)

Astrophotography began in 1845 when Jean Foucault and Armand Hippolyte Fizeau, in Paris, obtained the first daguerreotype of the Sun. Imaging the stars was more problematic. The main drawback of the daguerreotype process (other than its use of toxic mercury fumes in development) was its "slowness"; even in bright daylight, exposures of ten to fifteen minutes were required to record images. Thus, it was not until the night of July 16, 1850, that the first daguerreotype of a star—Vega—was taken by Harvard astronomers J. A. Whipple and William Cranch Bond.

Photography of the heavens is one area of astronomy in which amateurs have made, and continue to make, significant contributions. In 1852, amateur astronomer Warren De la Rue, in England, began to photograph celestial objects through a clock-driven reflector telescope, developing the plates in the dark space below the floor of his observatory. De la Rue's detailed images of the lunar surface earned him the Royal Astronomical Society's Gold Medal, and his daily photographs of the Sun confirmed the cyclical occurrence of sunspots.

The new astrophotography received a simultaneous boost from another amateur astronomer, this one a socially prominent New York lawyer. Lewis Morris Rutherfurd had taken up astronomy as a hobby while an undergraduate at Amherst. In 1856, he closed his law practice and built a large domed observatory in the garden of his Manhattan residence at 175 Second Avenue, near Eleventh Street. There he installed a transit telescope and an eleven-inch-wide refractor (later replaced by a thirteen-inch) that he constructed with his friend, telescope maker Henry Fitz.

Rutherfurd was put off by the slow daguerreotype process. Instead, he adopted the equally complex, but more light-sensitive, wet-collodion method invented by English sculptor Frederick Scott Archer in 1850. Prior to each exposure, Rutherfurd coated a glass plate with a solution of guncotton and various iodine and bromine compounds. (Guncotton, or nitrocellulose, is a highly volatile substance made by treating purified cotton with nitric and sulfuric acids and was used in the manufacture of explosives.) Rutherfurd allowed the moistened plate to dry before dipping it into a solution of silver salts. The prepared plate was sensitive to light only as long as it remained wet, a period of about six minutes, so it had to be loaded into the camera immediately and the exposure taken.

Rutherfurd began his photographic survey of the heavens with bright objects: the Sun, Moon, and planets. But soon he was recording the fields surrounding certain stars, such as 61 Cygni, in the belief that his plates might be used for parallax measurement. He constructed his own measuring engine for the purpose, yet never completed the project. By 1877, when he retired from observing, Rutherfurd had amassed more than 1,400 celestial photographs. He donated his telescopes and measuring engine to Columbia University in 1883, followed in 1890 by his extensive plate collection. The parallax determination of 61 Cygni was taken up by Herman Davis, who published the results in 1898.

Astronomer Charles Pritchard, at Oxford, joined the stellar parallax effort in 1886. Pritchard used highly sensitive dry-plate photography, developed in 1871 by English physician Richard Maddox. Dry plates came precoated with a silver-bromide gelatin and rendered the wet-collodion process obsolete. Having obtained the original reflector telescope of pioneer astrophotographer Warren De la Rue, Pritchard took 330 exposures of 61 Cygni over a two-year period. Although his measured parallax of 0.45 arcsecond was nearly 50 percent larger than Bessel's estimate, Pritchard is credited with demonstrating the viability of photographic parallax measurement.

Just as Friedrich Bessel had established standards for the use of the heliometer in stellar parallax measurement, American astronomer Frank Schlesinger established the corresponding standards for photography. Schlesinger characterized the stellar parallax quest as "a continual struggle between the necessities of the problem and the methods for attacking it, very similar to the conflict that has gone on between heavier and heavier artillery and stronger and stronger armour-plate." From 1903 to 1905, working at the Yerkes Observatory outside Chicago, Schlesinger photographed numerous stellar parallax candidates through the forty-inch-diameter refractor, the largest refractor telescope in the world. He studied the influence on photographic position measurement of various phenomena, such as atmospheric refraction, telescope flexure, exposure time, plate warping, nonuniform photographic emulsions, and development procedures. Afterward, he organized an international collaboration to measure stellar parallaxes photographically, involving the Greenwich and Royal Cape Observatories, Yerkes, Dearborn Observatory in Illinois, Allegheny and Sproul in Pennsylvania, Mount Wilson in California, and Leander McCormick in Virginia. The collaborating observatories adopted Schlesinger's methodology, from the preparation of the photographic plates to their exposure, development, and measurement, and then through the process of data reduction. As a result, the number of reliable parallaxes grew dramatically. Schlesinger's 1924 *General Catalogue of Stellar Parallaxes,* which combined data from all of the collaborating observatories, listed measurements for 1,870 stars. (Some 500,000 exposures have been taken at Allegheny Observatory alone.) The catalogue's original edition has since been succeeded by three others, the most recent, published by the Yale University Observatory in 1995, containing parallaxes of 8,112 stars.

The advent in the 1960s of large-aperture, high-precision reflector telescopes, such as the sixty-one-inch U.S. Naval Observatory instrument in Flagstaff, Arizona, allowed astronomers to measure parallaxes of very faint stars, a population that had largely been missed by previous photographic surveys. In its first twenty years of operation, over 50,000 photographic plates were taken with the Naval Observatory telescope, and some 800 new parallaxes were derived. The 1960s also saw the development of automatic measuring engines, which sweep unassisted over a photographic plate and are now able to determine the relative positions of star images at the rate of thousands every hour.

More recently, technology has once again revolutionized stellar-parallax detection. Silicon-based electronic components with light-sensitive surfaces have supplanted the photographic plate. These charged-coupled devices, or CCDs, are cousins of the light-detecting "chip" inside a home video camera. Now star images are viewed on a computer screen instead of on a backlit photographic plate. The darkroom and its chemicals have been eliminated. And parallax precision has improved almost tenfold compared to the best photographic determinations. (CCDs have not entirely displaced photographic technology in parallax measurement; relative to photography's broad field of view, the tiny light-sensitive chips are limited to only a tiny swath of sky at one time.) In CCD-based imaging devices, no longer is starlight energy "locked" within the chemical emulsion on a glass plate. With the star's photons transformed into their electronic avatars, the information once carried by these photons can be further transformed into radio signals and conveniently beamed, say, from observatory to office. Or, more importantly, from a space satellite to the ground.

The greatest hindrance to ground-based parallax measurement is the Earth's atmosphere. Passing through the ocean of air molecules that envelops our planet, photons from a distant star scatter like marbles falling through a Pachinko game. By the time they arrive at the telescope, they have been so buffeted from their original paths that the starry image they form in the eye or on the photographic plate is smeared out. The true position of the star, and hence its parallax, becomes difficult to pinpoint. Thus, the newest frontier for parallax research is outer space, above the distorting influences of the Earth's atmosphere. (An orbiting telescope also suffers none of the gravity-induced distortions of a telescope on the ground.)

The first spaced-based parallax observatory was proposed at a 1967 meeting of the International Astronomical Union in Prague and funded by the European Space Agency (ESA) in 1980. The satellite was designated Hipparcos, for *Hi*gh *P*recision *Par*allax *C*ollecting *S*atellite, in honor of the astronomer Hipparchus, who is alleged to have compiled the first "high-precision" star catalogue some 2,000 years ago. Hipparcos was launched in August 1989 from Kourou, French Guyana, atop an ESA Ariane-4 rocket. A malfunction in the booster's final stage left the 1.4-ton satellite short of its intended 22,000-mile-high circular orbit. Such a *geostationary* orbit had been designed to "park" the satellite above its base station, and would have permitted data to stream into researchers' computers twenty-four hours a day; instead, Hipparcos swung in a long, elliptical path that provided only an intermittent link with the ground. In an all-out effort to salvage the multimillion-dollar mission, ESA swiftly coordinated a worldwide array of receiving antennae and modified its own operational procedures to cope with Hipparcos's swooping orbit. By the time Hipparcos ceased functioning four years later in August 1993, every one of its original scientific objectives had been surpassed.

The light of the target stars was sufficiently intense that there had been no need to make the telescope itself large; indeed, Hipparcos's light-collecting mirror was eleven inches in diameter, only slightly wider than the mirror in a common amateur's telescope. (By comparison, the mirror in the Hubble Space Telescope is 94.5 inches across.) Hipparcos rotated slowly as it progressed around the Earth, sweeping its modest aperture across the heavens like a robotic William Herschel. It observed each target star about 100 times during its four-year lifetime. The raw data was radioed back to Earth and fed into banks of computers in what one mission scientist called the "biggest computation in the history of astronomy." The final product: a pair of stellar-parallax catalogues whose scope would have astonished parallax hunters of times past.

The main Hipparcos catalogue lists parallaxes (along with positions, proper motions, brightnesses, and colors) of 118,218 stars. The average precision of the parallaxes is about $^1/_{1,000}$ of an arcsecond, a couple of hundred times better than Bessel's heliometer measurements in the 1830s and equivalent to reading from Los Angeles a one-inch-high newspaper headline in New York or making out a sunbather on the Moon. Overall, Hipparcos has tripled the parallax astronomer's reach into space, to more than 300 light-years. Before Hipparcos, fewer than 1,000 star distances had been measured to better than 10 percent accuracy; now that number exceeds 22,000. Distances of the Sun's immediate neighbors—stars within thirty light-years—are trustworthy to an unprecedented 1 percent. Alpha Centauri, the nearest star beyond the solar system and the one observed by Thomas Henderson during his time in the "Dismal Swamp," has been pinpointed by Hipparcos at 4.395 light-years. Vega, once the target of the great Fraunhofer refractor of Wilhelm Struve, lies 25.3 light-years from Earth. And Friedrich Bessel's 61 Cygni gleams at us from a distance of 11.4 light-years, not far from where Bessel had placed it in 1838.

For the first time, the distance to the nearest major star cluster, the Hyades, has been determined by direct parallax measurement: 151 light-years. And the well-known Pleiades star cluster (the "Seven Sisters"), at 385 light-years, is some 10 percent closer than had previously been believed, forcing a reevaluation of theories of stellar energy production.

While Hipparcos's main telescope was surveying the sky, the satellite's guidance system—its so-called star-trackers—were simultaneously collecting star-position data. From the star-tracker results, mission scientists assembled the Tycho catalogue, named in honor of the sixteenth-century Danish astronomer. The Tycho catalogue contains lower-precision parallaxes of the brightest 2.5 *million* stars in the heavens. Clearly, the stellar parallax "problem" is no longer one of measurement—indeed with Hipparcos, the measurement process was almost mechanical—but of how to handle the abundance of data. And with a new generation of parallax satellites in the works, that "problem" is going to grow.

NASA, ESA, and a consortium of German universities are all planning their own Hipparcos successors. The German-made Double Interferometer for Visual Astronomy (DIVA) will be placed into orbit in 2004 and will measure the parallaxes of 30 million stars out to 3,000 light-years. NASA's Full-sky Astrometric Mapping Explorer (FAME) had been scheduled for launch in 2004, prior to its recent cancellation for budgetary reasons. It was to have measured positions, parallaxes, and proper motions of 40 million stars as far as 8,000 light-years, with accuracies up to twenty times better than Hipparcos. NASA's Space Interferometry Mission (SIM), in 2009, will observe a "mere" 20,000 stars, but with even greater precision than FAME. ESA's next entry into the stellar-parallax arena is the Global Astrometric Interferometer for Astrophysics (GAIA), slated for a 2010 launch. GAIA will measure positions and parallaxes of over a *billion* stars, or just about every star within reach of a moderately large, ground-based telescope. GAIA's precision is expected to be about 5 millionths of an arc-second, well over 100 times the acuity of Hipparcos. With such vision, you could make out a one-inch newspaper headline from the Moon or a flea from over 200 miles.

Astronomers had once sought the parallaxes of stars for a single reason: to validate the Sun-centered model of the cosmos. Today they strive to measure stellar parallaxes, and their associated distances, to solve a host of fundamental problems in astronomy. Here is but a partial list of how contemporary astronomers use their newfound abundance of parallax data:

- To more accurately calibrate the light output of "standard" stars, upon which rests the entire framework of cosmic distance indicators (By various means, astronomers can infer distances to celestial objects that lie beyond the range of parallax detection. However, nearly all such methods are calibrated from parallax-based stellar distances. The greater the number of accurate parallaxes, the more trustworthy the results of these "leveraged" methods of distance measurement.)
- To determine the true light output of stars in order to evaluate numerical models of stellar energy production, then to use these models to infer the distances, ages, and stellar composition of remote star clusters
- To derive the masses of binary-star systems, the most reliable indicator of stellar mass
- To deduce the true dimensions of nearby giant stars (in conjunction with ground-based observations)

- To assess the overall distribution of star types within the Galaxy in order to learn about star-formation rates and characteristics
- To determine the true light output and ages of newly formed stars
- To ascertain whether certain stars are members of a given star cluster or are merely chance alignments
- To infer overall patterns of movement of stars and subsequently the mass distribution within our region of the Galaxy (in conjunction with proper motion measurements)
- To deduce the age of the oldest star clusters, which sets a lower limit to the age of the universe
- To gauge more precisely how dust particles in outer space impede the transmission of starlight

The legacy of the early parallax hunters—starting with the first successful measurement of the distance to a star—permeates the present-day investigation of the cosmos. Indeed, Michael Perryman, former project scientist for the Hipparcos mission, has written, "Almost everything in astronomy depends in some way on knowing star distances."

Over the centuries, the study of stellar parallax has played an integral role in the art of astronomical observation and has served as a benchmark against which to gauge the astronomer's instruments, skill, and patience. The quest for stellar parallax began when Aristarchus set the Earth in orbit about the Sun, a motion that was "revolutionary" in both the scientific and historical senses of the word. The absence of stellar parallax was invoked against the heliocentric model by Hipparchus and Ptolemy, who nonetheless laid the footings for future observation and analysis of the heavens. In the 1500s, Copernicus revived the idea of the Sun-centered cosmos, fortifying his logical building blocks with the firm mortar of mathematics. Once again, stellar parallax became the pivotal factor in the clash between competing world systems. Copernicus's contemporary, Tycho Brahe, took up the search for the elusive parallax, but even his sharp eye and sophisticated instruments proved unequal to the task.

With his telescope, Galileo gave seventeenth-century astronomers unprecedented power to peer into space and to reveal previously hidden secrets of the universe. Even so, the stars appeared unmoved as the Earth completed its yearly circuit around the Sun. Robert Hooke attempted to bring technical sophistication to the enterprise, but he was defeated by the poor state of technology, as well as by his own impatience. Through very "un-Hooke-like" perseverance, James Bradley formulated the techniques of precision astronomy in the eighteenth century. Still, the parallaxes of stars escaped detection— although with his unexpected discovery of the aberration of starlight, Bradley at last proved the Copernican thesis that the Earth orbits the Sun. William Herschel coupled mammoth telescopes with mammoth ambition and attempted the wholesale measurement of stellar parallaxes, resurrecting a double-star method of Galileo's. Instead, Herschel's observations overthrew his own assumptions about double stars and left the issue of stellar distances unresolved.

The essential leap in telescopic sophistication came in the 1820s with the precision instruments of Joseph Fraunhofer and other German craftsmen. Two of Fraunhofer's state-of-the-art telescopes found their way into the able hands of Friedrich Bessel and Wilhelm Struve, while at the same time down in South Africa, Thomas Henderson struggled to wring honest results from a hand-me-down telescope he knew was flawed. In 1838 and 1839, three parallaxes were put before the scientific community, the first such measurements to carry any weight among skeptical astronomers. The pace of parallax discovery, initially languid, accelerated when the camera replaced the astronomer's eye at the back end of the telescope. And today, with one parallax satellite having been lofted into space and others poised to follow, astronomers are able to measure a million stellar parallaxes in the same time it took Friedrich Bessel to measure just one.

The parallax quest is a chronicle of human striving toward a common goal. It refutes the misperception that science is a steady forward march to enlightenment (a misperception, by the way, fostered by many science textbooks that, for the sake of brevity, highlight only scientific advancement). Science, as exemplified by the long hunt for stellar parallax, is rife with failure, false assumptions, unfounded optimism, jealousy, egotism—in short, science is a truly human enterprise. Look beyond its numbers and its logic and its supposedly objective façade, and you will discover (as I have, to my delight) the imperfect, yet inspiring, characters who populate the unvarnished story of scientific achievement.

Through the colorful lives I have lately come to know, my

own view of the parallax tale has acquired a new dimension: a human dimension, as revealing in its own way as the third dimension the parallax hunters themselves struggled to establish in the cosmos. The portraits I've collected for this book, once almost interchangeable in my mind, now resonate with individual accounts of frustration, failed dreams, and, through it all, the longing for success. In the centuries-long pursuit of the first stellar parallax, these astronomers are not just pioneers, but heroes, if only for facing down such daunting odds, for persevering where others might have given up. It is to them that modern astronomers owe a debt of gratitude. In this age of technological wonders, where computer screens glow and satellites soar and the accumulated knowledge of centuries can be accessed at a finger's touch, it is easy to forget how we got to where we are. Yet as the Chinese proverb wisely counsels, "When you drink from the well, remember the people who dug the well."

Acknowledgments ★★★

P*arallax* was written during an academic sabbatical granted to me by the University of Massachusetts Dartmouth, where I have taught astronomy and physics for more than twenty years. I am indebted to my colleagues in the Physics Department for taking up a variety of burdens so I could complete this book. My thanks go to the Center for Astrophysics at Harvard University for their hospitality during my sabbatical year and to the staff of the Wolbach Library, who were ever helpful in retrieving obscure books and journals.

My agent Sally Brady served as first reader of every version of every chapter; her high editorial standards and unremitting support drew out the writer in me and gave me the confidence to put my own personal stamp on *Parallax.* Erika Goldman, my editor at W. H. Freeman, taught me early on that the most compelling stories are the ones about people, a lesson that effectively made this book what it is. Science historian Owen Gingerich generously gave of his time to read the entire manuscript and point out historical inaccuracies and other errors in the text; he also provided several illustrations from his personal collection. My longtime friend Jane Langton, who has reviewed and encouraged my writing for many years, has been a constant inspiration. Ken Mallory, my neighbor and Director of Publications at the New England Aquarium, has followed the development of this book from its inception and has helped me forge valuable connections in the science-writing world.

My thanks also go to many others involved in the making of *Parallax:* Minnie Tai, editorial assistant at W. H. Freeman, for so cheerfully fielding my many e-mailed requests and questions; Roger Sinnott at *Sky & Telescope* magazine for his valuable insight into a variety of astronomical issues; Dorrit Hoffleit and William van Altena at Yale for suggesting relevant references and for providing information about the Yale heliometer; the librarians at Countway Library, Harvard Medical School, for their assistance in researching the invention of spectacles; Peter Hingley at the Royal Astronomical Society Library for his expertise about sources of historical images in astronomy; Margaret Hirshfeld and Wolfhard Kern for helping me polish my rusty German; Ellen Shapiro and Roberto Pietroforte for translating a sample of Galileo's "colorful" Italian; Scott Ransom at Harvard and S. Duckworth at the U.S. Army Field Artillery School for providing information about military range finders; Toby Huff for a concentrated lesson on Islamic astronomy; and Ken Yee for providing the Chinese proverb that closes this book. The following people helped me track down the historical images that appear in *Parallax:* Alan Batten, Dominion Astrophysical Observatory; Elizabeth Danne, Bonn University Observatory; Richard Dreiser, Yerkes Observatory; Irene Falnov, Fredericksborg Castle Museum; Paolo Galluzi and Franca Principe, Museum of Science, Florence; Jane Holmquist, Princeton University; Ethleen Lastovica, South African Astronomical Observatory; Laurits Leedjarv, Tartu Observatory, Estonia; Derek Ohland, South African Museum; Margrit Prussat, Deutsches Museum; and Taavi Tuvikene, Tartu University.

Two final acknowledgments: to my wife Sasha, for the unflagging patience and support that allowed me to fulfill this longtime dream; and to my aunt, Alice Popper, whose gift of an astronomy book when I was a kid set off a forty-year sequence of events that culminated in *Parallax*.

Notes and Further Reading

STELLAR PARALLAX THROUGH THE AGES				
Observer or technique	Date	Relative measurement precision (smaller is better)	Distance range (light-years)	Number of parallax stars added
Hipparchus	150 B.C.	300,000,000		0
Tycho Brahe	1600	60,000,000		0
John Flamsteed	1700	10,000,000		0
James Bradley	1750	500,000		0
Bessel/Struve/ Henderson	1839	100,000	25	3
Heliometer	1900	100,000	25	Less than 100
Photography	1950	50,000	100	Several thousand
Photography	1970	5,000	100	Several thousand
Hipparcos (ESA)	1989	1,000–20,000	300	2.5 million
DIVA (Germany)	2003	200–5,000	3,000	30 million
FAME (NASA)	2004	50–500	8,000	40 million
SIM (NASA)	2005	3	100,000	20,000
GAIA (ESA)	2009	3–200	30,000	1 billion

Sources: Hoffleit (1949); Kovalevsky (1990); European Space Agency, NASA, and DIVA web sites, as follows:
ESA: astro.estec.esa.nl/Hipparcos
         astro.estec.esa.nl./SA-general/Projects/GAIA/gaia.html.
NASA: sim.jpl.nasa.gov/index.html
         aa.usno.navy.mil/FAME
DIVA: www.aip.de/groups/DIVA/diva.html

Chapter 1: REINVENTING THE COSMOS

p. 2, *included a hundred-and-eighty-foot golden phallus:* Stille (2000), p. 90.

p. 2, *Into this percolating cauldron of ideas:* The Egyptian government and the United Nations Educational, Scientific and Cultural Organization (UNESCO) are jointly building a new Alexandrian library. The project received an unintended boost during a 1974 visit to Egypt by Richard Nixon, when the former president asked to see the site of the ancient Library. His Egyptian hosts confessed in embarrassment that they hadn't a clue where it was. Stille (2000).

p. 3, *In our endeavor to understand reality:* Einstein and Infeld (1938), p. 33.

p. 4, *Observation . . . is the pitiless critic of theory:* Clerke (1902), p. 2.

p. 10, *When the Roman legions attacked:* James and Thorpe (1994), p. 220.

p. 10, *such terror had seized upon the Romans:* Dryden (1957), p. 378.

p. 10, *Give me where to stand:* Heath (1990), p. 399.

p. 11, *Archimedes was to be believed:* Heath (1990), p. 399.

p. 11, *not as matters of any importance:* Dryden (1957), p. 376.

p. 11, *forget his food and neglect his person:* Dryden (1957), p. 378.

p. 12, *I will try to show:* Heath (1990), p. 520.

p. 14, *There was a young fellow from Trinity:* Gamow (1961), p. i. Reprinted by permission, Dover Publications Inc. In the original, the phrase "the square root of infinity" was expressed in mathematical symbols.

Chapter 2: THE CIRCLE GAME

p. 22, *Hipparchus . . . detected a new star:* Sarton (1959), p. 301.

p. 24, *Ptolemy's attempt to write:* Gingerich (1993b), p. 5.

p. 30, *When a teacher in a Faculty of Arts:* Hoskin (1997), p. 74.

Chapter 3: WHAT IF THE SUN BE CENTER TO THE WORLD?

p. 34, *their velocities so accurately adjusted:* From a report by David Rittenhouse to the American Philosophical Society, as quoted in Babb (1998). See also Leitch (1978).

p. 36, *The lurking demons:* Thiel (1957), p. 75.

p. 36, *tornado of a man:* Hoyle (1973), p. 26.

p. 38, *most remote corner of the Earth:* Gingerich (1993b), p. 164.

p. 38, *800 building stones:* Rosen (1971), p. 342.

p. 38, *[Past astronomers] have not been able:* From *De Revolutionibus Orbium Coelestium*, translation by Hoskin (1997), p. 92.

p. 39, *What appear to us as motions of the Sun:* Rosen (1971), p. 339.

p. 39, *neither sufficiently perfect:* Gingerich (1993b), p. 28.

p. 40, *I know of a modern scientist:* Gingerich (1993b), p. 222, from the preface of Reinhold's commentary on Peurbach's *Theoricae novae planetarum*.

p. 41, *the man he referred to as Domine Praeceptor:* Koestler (1989), p. 159.

p. 41, *You come to see me at the same age:* Koestler (1989), p. 193.

p. 41, *The Ptolemaic astronomy is nothing:* Rosen (1971), p. 194.

p. 41, *linked most nobly together:* Gingerich (1993b), p. 166.

p. 41, *The astronomer who studies the motions:* Koestler (1989), p. 165.

p. 42, *[T]he scorn which I had reason to fear:* Rosen (1978), p. 3.

p. 42, *Perhaps there will be babblers:* Rosen (1978), p. 5.

p. 42, *[I]t is not necessary for the [book's] hypotheses:* 1994 translation of *De Revolutionibus* by Owen Gingerich.

p. 43, *It seems to me that your Reverence:* Gingerich (1993b), p. 274.

p. 43, *modern-equivalent price of over $100:* Gingerich (1993b), p. 262.

p. 43, *turn the whole art of astronomy upside down:* Kopal (1973), p. xi.

p. 43, *No one in his senses:* Boorstin (1983), p. 319.

p. 44, *Here Copernicus is dreaming:* Gingerich (1992), p. 73.

p. 44, *willful destruction of a De Revolutionibus:* Gingerich (1992), p. 81.

p. 44, *In the center of all resides the Sun:* From *De Revolutionibus,* in Hoskin (1997), p. 97.

p. 44, *so linked together that in no portion:* Rosen (1978), p. 5.

p. 44, *What a picture:* Motz and Weaver (1995), p. 62.

p. 45, *From Saturn, the highest of the planets:* Rosen (1978), p. 22.

Chapter 4: CROSSED EYES AND WOBBLING STARS

p. 50, *Virgil told me later:* Sacks (1995), p. 114.

p. 51, *are not given the world:* Sacks (1995), p. 114.

p. 51, *vision is pure sensation:* Dillard (1974), p. 45.

p. 51, *I'm told I reached for the moon:* Dillard (1974), p. 45.

p. 61, *wrapped in the cloak of authority:* Van Helden (1985), p. 3.

p. 62, *One can speak lightly of going to St. Helena:* Evans, D. S. (1988), p. 9.

p. 65, *There will be no other [transit of Venus]:* Ashbrook (1984), p. 230.

Chapter 5: THE HEAVENS ERUPT

p. 78, *He had Tycho educated by private tutors:* During the Middle Ages, it was not unusual for students to enter college at age twelve or thirteen. But by 1559, according to a study by historian Owen Gingerich, the median age of matriculation had risen to around eighteen. Tycho was indeed a prodigy.

p. 80, *a narrow, curved mark:* Ashbrook (1984), p. 5.

p. 82, *Behold, directly overhead:* Clark and Stephenson (1977), p. 174.

p. 84, *this star is not some kind of comet:* From *De Nova Stella,* quoted in Shapley and Howarth (1929), p. 19.

p. 84, *formed by the ascending from earth:* Dreyer (1890), p. 65.

p. 85, *common herd of scribblers:* Dreyer (1890), p. 57.

p. 85, *An astronomer . . . more than the students:* Christianson, J. (1961), p. 127.

p. 85, *would lose its greatest ornament:* Ley (1963), p. 87.

p. 86, *Curiously, the names "Rosenkrans" and "Guldensteren":* Gingerich (1981), p. 395.

p. 88, *chattered incessantly:* Dreyer (1890), p. 128.

p. 88, *The squire is on the land:* Dreyer (1890), p. 128.

p. 88, *Tycho was an oppressive landlord:* Ashbrook (1984), p. 3.

p. 89, *a six-year-old Johannes Kepler:* Koestler (1989), p. 234.

p. 90, *When [Tycho] measured the altitude of a star:* Ley (1963), p. 89.

p. 90, *It is a difficult matter:* Shapley and Howarth (1929), p. 14.

p. 91, *I now no longer approve of the reality:* Hartner (1969), p. 37.

p. 92, *pulled down and renovated:* Dreyer (1890), p. 376.

p. 92, *Tycho was an aristocrat:* Koestler (1989), p. 306.

p. 93, *obstructed in his progress:* Koestler (1989), p. 308.

p. 93, *Let me not seem to have lived in vain:* Dreyer (1890), p. 386.

Chapter 6: THE TURBULENT LENS

p. 97, *An exhaustive modern-day archival search:* Ronchi (1967), p. 125.

p. 97, *The purpose of sight is to know the truth:* Ronchi (1967), p. 125.

p. 98, *Much has been written:* Rivista di oftalmologia 1 (1946), p. 140, as quoted in Rosen (1956), p. 13.

p. 99, *They are badly polished:* Rosenthal (1996), p. 492.

p. 100, *"supernormal vision":* See Miller (2000), pp. 31–36.

p. 101, *Historian Albert van Helden:* van Helden (1977), p. 10.

p. 102, *a lump of opaque blue vitreous paste:* Kurinsky (1991), p. 3.

p. 103, *The room where the raw materials:* Ellis (1998), p. 19.

p. 105, *the appearance of the ingredients:* Kurinsky (1991), p. 42.

p. 106, *tipping point:* Gladwell (2000).

p. 109, *Thus from an incredible distance we might read:* Bacon, Roger. *Opus Majus,* circa 1267, quoted in van Helden (1977), p. 28.

p. 109, *a certain device by means of which:* van Helden (1977), p. 20.

p. 110, *the said glasses are very useful in sieges:* van Helden (1985), p. 64.

Chapter 7: THE WRANGLER OF PISA

p. 114, *It appears to me that they:* Drake (1992), p. 23.

p. 115, *[B]efore answering the opposing reasons:* Koestler (1989), p. 458.

p. 115, *ridiculoso, elefantissimo:* Ley (1963), p. 126.

p. 115, *The picture of Galileo's personality:* Drake (1992), p. 3.

p. 117, *adopted the teaching of Copernicus:* Koestler (1989), p. 361.

p. 117, *Italian whose last name was the same as his first:* Koestler (1989), p. 361.

p. 117, *Stand forth, O Galileo:* Gingerich (1993b), p. 394.

p. 117, *Even if we could detect no displacement:* Koestler (1989), p. 364.

p. 120, *There have been numerous gentlemen:* Galilei (1989), Introduction, p. 6.

p. 122, *uneven, rough, and crowded:* Galilei (1989), p. 40.

p. 122, *It is like the face of the Earth itself:* Galilei (1989), p. 40.

p. 122, *To whatever region of [the sky]:* Galilei (1989), p. 62.

p. 123, *In this short treatise:* Galilei (1989), p. 35.

p. 123, *complete their great revolutions:* Galilei (1989), p. 31.

p. 124, *Behold, therefore, four stars:* Galilei (1989), p. 31.

p. 124, *The lens, with a crack running its width:* Galileo's lens can be viewed on the museum's Web site [http://galileo.imss.firenze.it/museo/4/eiv01.html]

p. 124, *According to a letter Galileo sent to Cosimo:* Galilei (1989), p. 20.

p. 126, *the triumph "of the great Copernican system":* Galilei (1964), Introduction, p. xxii.

p. 127, *They have endeavored to spread the opinion:* Drake (1992), p. 6.

p. 127, *the story of the mind of Signor Galileo:* Galilei (1964), Introduction, p. xxx.

p. 128, *[H]e had gambled everything:* Galilei (1964), Introduction, p. xxxviii.

p. 129, *"This universe," he wrote to a friend:* Sobel (1999), p. 355.

p. 129, *I, with you, would say that:* Galilei (1964), p. 397.

p. 130, *make use of instruments:* Galilei (1964), p. 398.

p. 131, *I do not think that the stars are spread:* Galilei (1964), p. 393.

Chapter 8: THE ARCHIMEDEAN ENGINE

p. 138, *wild frightfull dreames:* 'Espinasse (1956), p. 153.

p. 138, *of any man in the world that I ever saw:* Andrade (1950), p. 440.

p. 139, *but of middling stature:* Dick (1957), p. 165.

p. 139, *uncut and lank:* Andrade (1950), p. 440.

p. 139, *Apples agreed well:* Nichols (1999), p. 122.

p. 139, *dissolve that viscous Slime:* 'Espinasse (1956), p. 152.

p. 139, *Hooke's spiny character:* Westfall, "Robert Hooke," in Gillespie (1972).

p. 139, *He went stooping and very fast:* Keynes (1960), p. vi.

p. 140, *an ignorant impudent ass:* Diary, November 27, 1677; Nichols (1999), p. 120.

p. 140, *much troubled with Mr. Hooke:* Chapman (1990), p. 51.

p. 140, *invented thirty severall wayes of Flying:* Dick (1957), p. 165.

p. 140, *to furnish the Society every day they meete:* 'Espinasse (1956), p. 4.

p. 140, *Mr. Hooke proposed:* Keynes (1960), p. ix.

p. 141, *the most ingenious book:* 'Espinasse (1956), p. 58.

p. 142, *[He] steales flowers from others:* 'Espinasse (1956), p. 7.

p. 142, *Shewd my quadrant to all but Oldenburg:* 'Espinasse (1956), p. 106.

p. 142, *By thinking of them without ceasing:* Ferris (1988), p. 116.

p. 143, *Hooke sniffed that the little lens:* Christianson, G. (1987), p. 20.

p. 143, *Two such as each seem'd worthiest:* 'Espinasse (1956), p. 1.

p. 143, *the Greatest Discovery in Nature:* Dick (1957), p. 167.

p. 144, *Spent most of my time in considering all matters:* 'Espinasse (1956), p. 6.

p. 144, *Whether the Earth move or stand still:* Hooke (1674), p. 1.

p. 144, *somewhat extravagant, and hardly practicable:* Hooke (1674), p. 10.

p. 145, *be the Air thicker or thinner:* Hooke (1674), p. 15.

p. 145, *conversant only with illiterate persons:* Hooke (1674), p. 2.

p. 147, *Inconvenient weather and great indisposition:* Hooke (1674), p. 24.

p. 147, *at a Gentlemans house:* Hooke (1674), p. 23.

p. 148, *because of the damp of the place:* Howse (1995), p. 58; Francis Place's etching of Flamsteed's well telescope is reproduced in the book as Plate XIIa.

p. 148, *An unhappy accident:* Hooke (1674), p. 24.

p. 148, *'Tis manifest then by the observations:* Hooke (1674), p. 25.

Chapter 9: A COAL CELLAR WITH A VIEW

p. 152, *The slightest inconsistency:* Clerke (1902), p. 3.

p. 152, *I think we shall be inclined to admit:* Turner (1963), p. 87.

p. 152, *peculiarly kind and gentle:* Rigaud (1832), p. ciii.

p. 152, *character in every respect:* Turner (1963), p. 113.

p. 152, *the best astronomer in Europe:* Turner (1963), p. 113.

p. 153, *averse from the promiscuous conversation:* Rigaud (1832), p. ciii.

p. 153, *his record indeed was such:* Turner (1963), p. 90.

p. 154, *the best mechanician of his time:* King (1955), p. 110.

p. 156, *therefore curiosity that tempted me:* Rigaud (1832), p. 2.

p. 157, *a rainy, blowing tempestuous night:* Stewart (1964), p. 103.

p. 158, *before he would rely on it:* Chapman (1990), p. 88.

p. 158, *were always making discoveries:* Roberts (1989), p. ix.

p. 158, *At last, when [Bradley] despaired:* Stewart (1964), p. 105.

p. 164, *unremitting toil:* Turner (1963), p. 87.

p. 168, *It must be granted:* Shapley and Howarth (1929), p. 108.

p. 168, *stars "of greatest lustre":* Woolley (1963), p. 50.

Chapter 10: DOUBLE VISION

p. 172, *shabby curate who has strayed:* Auden, W. H. 1962. *The Dyer's Hand and Other Essays.* New York: Random House.

p. 173, *A knowledge of the construction:* Herschel, W. (1912), Vol. II, p. 459.

p. 173, *resolved to examine every star:* Herschel, W. (1782), p. 97.

p. 173, *laborious but delightful research:* Herschel, W. (1782), p. 89.

p. 174, *parade of blushing damsels:* Clerke (1895), p. 12.

p. 175, *to the tender mercies of her mother:* Lubbock (1933), p. 34.

p. 175, *thrown like balls on the shore:* Lubbock (1933), p. 48.

p. 175, *I never forgot the caution:* Lubbock (1933), p. 45.

p. 176, *She was attached during the 50 years:* Lubbock (1933), p. 58.

p. 176, *used to retire to bed:* Holden (1881), p. 36.

p. 176, *Bought a book of astronomy:* Lubbock (1933), p. 60.

p. 177, *[W]hen I read of the many charming discoveries:* Gingerich (1984a), p. 318.

p. 177, *knowledge indeed being very confined:* Ashbrook (1984), p. 128.

p. 178, *One early biographer mistakenly reported:* Lubbock (1933), p. 63.

p. 178, *A cabinet maker making a tube:* Lubbock (1933), p. 65.

p. 178, *resembled an astronomer's:* Lubbock (1933), p. 73.

p. 178, *There it is at last:* Lubbock (1933), p. 73.

p. 178, *I was so much attached:* Gingerich (1984a), p. 318.

p. 178, *an immense quantity:* Lubbock (1993), p. 89.

p. 178, *served me for many hours' exercise:* Lubbock (1993), p. 89.

p. 179, *Both my Brothers:* Lubbock (1933), p. 90.

p. 179, *Double stars which they could not see:* Holden (1881), p. 65.

p. 179, *Seeing is in some respects:* Clerke (1895), p. 16.

p. 180, *perceived one that appeared visibly larger:* Herschel, W. (1781), p. 492.

p. 181, *It has generally been supposed:* Gingerich (1984a), p. 318, from an autobiographical letter Herschel wrote in 1784 for *The European Magazine.*

p. 181, *Among opticians and astronomers:* Lubbock (1933), p. 114, from a letter written by William Herschel to his sister Caroline, June 3, 1782.

p. 181, *Never bought Monarch honour so cheap:* Lubbock (1933), p. 133.

p. 181, *I employed myself:* Herschel, W. (1912), Vol. I, p. xxxvii.

p. 182, *I could give a pretty long list:* Lubbock (1933), p. 137.

p. 183, *The gardens and workrooms were swarming:* Sidgwick (1954), p. 134.

p. 184, *Come, my Lord Bishop:* Lubbock (1933), p. 157.

p. 184, *He seems a man without a wish:* Lubbock (1933), p. 169.

p. 184, *English lady-wife:* Lubbock (1933), p. 177, from a recollection by John Herschel's wife.

p. 184, *It was a mighty bewilderment:* Lubbock (1933), p. 168, from Holmes's *The Poet at the Breakfast-Table,* 1872.

p. 185, *[In] October I received £12,10:* Lubbock (1933), p. 172.

p. 185, *Would you think it proper:* Learner (1981), p. 61.

p. 185, *[If] satellites will come in the way:* Lubbock (1933), p. 165. The final postscript to Herschel's "Catalogue of a second Thousand of new Nebulae and Clusters of Stars" reads in full: "P.S. The planet Saturn has a sixth satellite revolving round it in about 22 hours, 48 minutes. Its orbit lies exactly in the plane of the Ring, and within that of the first satellite. An account of its discovery with the forty-feet reflector, and a more accurate determination of its revolution and distance from the planet will be presented to the Royal Society at their next meetings. WILLIAM HERSCHEL." (*Philosophical Transactions* 79 [1789], p. 255.) Herschel had glimpsed Enceladus two years earlier with his twenty-foot telescope, but apparently had been too busy to make confirming observations. Or perhaps he chose to credit the discovery to the forty-foot scope to reinforce the impression that the royal largesse had been well spent.

p. 186, *To find the distance of the fixed stars:* Herschel, W. (1782), p. 82.

p. 186, *[T]he whole diameter:* Herschel, W. (1782), p. 82.

p. 187, *so as to be able at last to say:* Herschel, W. (1782), p. 89.

p. 187, *guarded against optical delusions:* Herschel, W. (1782), p. 98.

p. 187, *not condemn his instrument:* Herschel, W. (1782), p. 100.

p. 188, *highly probable in particular:* Hoskin (1963), p. 37, from *Philosophical Transactions* 57 (1767), 234–264.

p. 188, *some of the great number of double, triple, stars:* Hoskin (1963), p. 37, from a letter dated May 26, 1783, published in *Philosophical Transactions* 74 (1784), 35–57.

p. 189, *I shall soon communicate:* Herschel, W. (1912), Vol. II, p. 205.

p. 189, *I want you to assist me:* Clerke (1895), p. 128.

p. 190, *I keep this as a relic:* Clerke (1895), p. 128.

p. 190, *You will see what a solitary and useless life:* King (1955), p. 142, from Sime (1900), p. 254.

p. 190, *if I was but thirty or forty years younger:* Lubbock (1933), p, 372.

Chapter 11: Dismal Swamp

p. 195, *Celestial maps had become:* Clerke (1895), p. 184, based on impressions of astronomer John Herschel around 1840.

p. 195, *his Majesty was very well pleased:* Dick (1957), p. 121.

p. 196, *When it is remembered:* Gill (1913), p. v.

p. 196, *There is a report here:* Warner (1995), p. 27.

p. 197, *Conceive yourself in a far distant land:* Warner (1995), p. 43.

p. 197, *spot near Mr. Coetsey's farmhouse:* Gill (1913), p. ix.

p. 198, *It is difficult to conceive [of] a man:* Gill (1913), p. xiv.

p. 198, *one day took offence at Mr. Fallows:* Warner (1995), p. 50.

p. 202, *insidious, venemous [sic] snakes:* Evans, D. S. (1988), p. 41.

p. 202, *lurking beside the bed:* Evans, D. S. (1988), p. 41.

p. 202, *Perhaps I may be pardoned:* Gill (1913), pp. xvi–xvii.

p. 202, *too honourable a man:* Gill (1913), p. xvi.

p. 203, *Henderson was rather the refined observer:* Gill (1913), p. xvii.

Chapter 12: The Twice-Built Telescope

p. 210, *Write to me now and then:* Wattenberg (1929b), p. 9.

p. 211, *One summer Saturday in 1804:* Bessel (1875), p. xviii.

p. 212, *I am delighted to report:* Wattenberg (1929b), p. 11.

p. 212, *happy and quiet:* Hamel (1984), p. 25.

p. 212, *Take care of your health:* Hamel (1984), p. 22.

p. 214, *once in the workshop of the artisan:* Bessel (1848), p. 17.

p. 216, *The French emperor is reported:* Gause (1968), p. 348.

p. 219, *In a letter to his mentor Olbers:* Williams, M. E. W. (1981), p. 123.

p. 219, *The history of annual parallax:* Fernie (1975), p. 226.

Chapter 13: QUEST FOR PRECISION

p. 226, *geometers and scientists . . . workmen:* King (1955), p. 229.

p. 226, *The fertility of this artist's genius:* King (1955), p. 230.

p. 231, *This is the man we have been looking for:* Roth (1976), p. 38.

p. 232, *Into the vacuum created by the English tax:* The glass tax was initially levied on windows. It is reported that Caroline Herschel became upset when she learned that her brother William had rented for their residence a thirty-window house. In the 1790s, William Pitt the Younger raised the excise tax and extended it to decorative glassware, chandeliers, and optical glass. The tax was not repealed until 1851. Learner (1981), p. 69.

Chapter 14: SO MANY GRASSHOPPERS

p. 239, *It was not the person who locked his invention:* Grant (1852), p. 533, as quoted in *Gentleman's Magazine* in 1790.

Chapter 15: THE STAR IN THE LYRE

p. 245, *We Struves cannot live happily:* Excerpt of a letter written by Jacob Struve. Batten (1988), p. 6.

p. 245, *The best kind of heroism:* Schickel, Richard. 1999. *Matinee Idylls: Reflections on the Movies.* Chicago: Ivan R. Dee.

p. 250, *Both mental and bodily fatigue:* Otto Struve, in Batten (1988), p. 16.

p. 251, *The opportunity to acquire this instrument:* Batten (1988), p. 46.

p. 252, *I stood before this beautiful instrument: Astronomische Nachrichten,* December 31, 1824, in Batten (1988), p. 47.

p. 252, *What a difference is seen:* Batten (1988), p. 47.

p. 254, *a magnificent work ranking among the greatest:* Gillespie (1972), p. 110.

p. 255, *by no means complete enough:* Dick and Ruben (1988), p. 119.

Chapter 16: THE SUBTLE WEAVE

p. 257, *The important thing, however:* Struve (1956), p. 10.

p. 258, *I well remember to have seen:* Herschel, J. (1842), p. 448.

p. 261, *I think Struve has taken the lead:* Struve (1956), p. 71.

p. 262, *secure [his] meaning from indistinctness:* Bessel (1838a), p. 152.

p. 262, *Having succeeded in obtaining:* Bessel (1838a), p. 152.

p. 262, *I have here troubled you:* Bessel (1838a), p. 161.

p. 262, *[T]he results have been placed:* Herschel, J. (1842), p. 443.

p. 264, *one of the greatest discoveries:* Wilhelm Struve, as quoted in Batten (1988), p. 127.

p. 264, *among the memories transmitted to me:* Struve (1956), p. 72.

p. 265, *I congratulate you and myself:* Herschel, J. (1842), p. 453.

p. 265, *The barrier has begun to yield:* Herschel, J. (1842), p. 453.

p. 266, *Nun wieder in die Sternwarte:* Letter dated September 14, 1843, quoted in Repsold (1919), p. 211.

p. 267, *The visibility of countless stars:* Kopal (1985), p. 7.

Epilogue: A DRINK FROM THE WELL

p. 269, *and thus we die, Still searching:* Robinson, Edwin Arlington. 1897. *The Children of the Night.* New York: Charles Scribner's Sons. Reprinted by permission, Simon & Schuster.

p. 270, *Brobdingnagian orbs:* Clerke (1905), p. 297.

p. 271, *certain idiosyncrasies of perception:* Clerke (1905), p. 291.

p. 272, *We have seen the views:* Leggat (2000), "The Daguerreotype."

p. 272, *People were afraid at first:* Leggat (2000), "The Daguerreotype."

p. 274, *a continual struggle:* Schlesinger (1927), as quoted in Shapley (1960), p. 113.

p. 276, *biggest computation in the history of astronomy:* European Space Agency's Hipparcos Web site [astro.estec.esa.nl/Hipparcos].

p. 279, *Almost everything in astronomy depends:* Perryman (1999), p. 42.

Abetti, Giorgio. 1952. *The History of Astronomy.* New York: Henry Schuman, Inc.

Aitken, R. G. 1936. *The Binary Stars.* New York: McGraw-Hill.

Allen, Richard H. 1963. *Star Names: Their Lore and Meaning.* New York: Dover Publications.

Ambronn, L. 1899. *Handbuch der Astronomischen Instrumentkunde.* Berlin: Springer Verlag.

Andrade, E. N. da C. 1950. "Wilkins Lecture: Robert Hooke." *Proceedings of the Royal Society* 201A, 439–473.

Ashbrook, Joseph. February 1963. "The 'Long Night' of Selenography." *Sky & Telescope* 29, 92–94.

———. August 1970. "Edmund Halley at St. Helena." *Sky & Telescope* 40, 86–87.

———. September 1975. "Old Greenwich Observatory and Flamsteed's Well." *Sky & Telescope* 50, 157.

———. 1984. *The Astronomical Scrapbook: Skywatchers, Pioneers, and Seekers in Astronomy.* Cambridge, MA: Sky Publishing Corporation.

Babb, M. I. 1998. "The Relation of David Rittenhouse and His Orrery to the University," University of Pennsylvania. [www.library.upenn.edu/vanpelt/pennhistory/orrery/orrery.html]

Baily, Francis. 1835. *An Account of the Rev. John Flamsteed.* London: William Clowes and Sons.

Batten, Alan H. 1988. *Resolute and Undertaking Characters: The Lives of Wilhelm and Otto Struve.* Dordrecht: D. Reidel.

Beckett, Francis, and Christensen, Charles. 1921. *Tycho Brahe's Uraniborg and Stjerneborg on the Island of Hveen.* London: Oxford University Press.

Bell, Louis. 1922. *The Telescope*. New York: McGraw-Hill.

Bessel, Friedrich Wilhelm. 1831. "Vorläufige Nachricht von einem auf der Königsberger Sternwarte befindlichen grossen Heliometer." *Astronomische Nachrichten* 8, no. 189, 396–426.

―――. November 9, 1838(a); December 14, 1838. "A Letter from Professor Bessel to Sir J. Herschel, Bart., Dated Königsberg, Oct. 23, 1838." *Monthly Notices of the Royal Astronomical Society* 4, no. 17, 152–161; no. 18, 163.

―――. December 13, 1838(b). "Bestimmung der Entfernung des 61sten Sterns des Schwans." *Astronomische Nachrichten* 16, 66–95.

―――. 1848. *Populäre Vorlesungen über wissentschafliche Gegenstände.* Schumacher, H. C., ed. Hamburg: Perthes-Besser and Mauke.

―――. 1875. *Abhandlung von Friedrich Wilhelm Bessel*. Engelmann, Rudolph, ed. Leipzig: Verlag von Wilhelm Engelmann.

Blackwell, D. E. 1963. "The Discovery of Stellar Aberration." *Quarterly Journal of the Royal Astronomical Society* 4, 44–46.

Blumenberg, Hans. 1987. *The Genesis of the Copernican World.* Cambridge, MA: Massachusetts Institute of Technology Press.

Boorstin, Daniel. 1983. *The Discoverers*. New York: Random House.

Both, Ernst E. July 1977. "Joseph von Fraunhofer." *Sky & Telescope* 54, 49–50.

Bradley, James. 1727–1728. "An Account of a New-Discovered Motion of the Fixed Stars." *Philosophical Transactions* 35, 636–661.

―――. 1748. "An Apparent Motion Observed in Some of the Fixed Stars." *Philosophical Transactions* 45, 1–43.

Brahe, Tycho. 1901. *Astronomiae Instauratae Mechanica* (Facsimile edition). Holmiae: P. A. Norstedt.

―――. 1913. *Opera Omnia* (includes *De Nova Stella*). Dreyer, J. L. E., ed. Copenhagen: Nielsen and Lydiche.

Brück, Hermann A. 1983. *The Story of Astronomy in Edinburgh.* Edinburgh: Edinburgh University Press.

Bukowski, Jerzy. April 1973. "In the Footsteps of Copernicus." *Courier,* 5–19.

Burmeister, Karl Heinz. 1967–1968. *Georg Joachim Rhetikus, 1514–1574: Eine Bio-bibliographie.* Wiesbaden: Pressler-Verlag.

Burnham, Robert, Jr. 1978. *Burnham's Celestial Handbook.* New York: Dover Publications.

Chance, W. H. S. 1937. "The Optical Glassworks at Benediktbeuern." *Proceedings of the Physical Society* (London) 49, part 5, no. 275, 433–443.

Chapman, Allan. 1990. *Dividing the Circle: The Development of Critical Angular Measurement in Astronomy, 1500–1850.* New York: Ellis Horwood.

―――. 1993. "The Astronomical Revolution." In *Möbius and His Band: Mathematics and Astronomy in Nineteenth-Century Germany.* Fauvel, John, et al., eds. Oxford: Oxford University Press, 35–76.

Christianson, Gale. July 1987. "Newton's *Principia:* A Retrospective." *Sky & Telescope* 74, 18–20.

Christianson, John. February 1961. "The Celestial Palace of Tycho Brahe." *Scientific American* 204, 118–128.

Church, John A. March 1963. "Optical Designs of Some Famous Refractors." *Sky & Telescope* 63, 302–308.

Clark, David H., and Stephenson, F. Richard. 1977. *The Historical Supernovae.* New York: Pergamon Press.

Clerke, Agnes. 1895. *The Herschels and Modern Astronomy.* New York: Macmillan and Company.

———. 1902. *A Popular History of Astronomy during the Nineteenth Century.* London: Adam & Charles Black.

———. 1905. *The System of the Stars,* 2nd ed. London: Adam & Charles Black.

Cohen, I. Bernard. January 1942. "The Astronomical Work of Galileo Galilei (1564–1642)." *Sky & Telescope* 1, 3–5.

Cole, K. C. 1999. *First You Build a Cloud.* New York: Harcourt Brace.

Comparato, Frank E. 1965. *Age of Great Guns.* Harrisburg, PA: Stackpole Company.

Cook, Alan. 1998. *Edmond Halley: Charting the Heavens and the Seas.* Oxford: Clarendon Press.

Copernicus, Nicolaus. 1944. *De Revolutionibus Orbium Coelestium* (Facsimile of original manuscript). Munich: Verlag R. Oldenbourg.

———. 1965. *De Revolutionibus Orbium Coelestium* (Facsimile of first edition, 1543). New York: Johnson Reprint Corporation.

Cowen, Ron. December 18/25, 1999. "Danish Astronomer Argues for a Changing Cosmos." *Science News* 156, vii.

Crowe, Michael J., ed. 1998. *A Calendar of the Correspondence of Sir John Herschel.* Cambridge, England: Cambridge University Press.

Dewhirst, D. W. 1955. "Observatories and Instrument Makers in the Eighteenth Century." *Vistas in Astronomy* 1, 139–143.

Dick, Oliver Lawson, ed. 1957. *Aubrey's Brief Lives.* Ann Arbor: University of Michigan Press.

Dick, W. R., and Ruben, G. 1988. "The First Successful Attempts to Determine Stellar Parallaxes in the Light of the Bessel/Struve Correspondence." In *Mapping the Sky: Past Heritage and Future Directions* (International Astronomical Union Symposium No. 133). Debarat, S., et al., eds. Dordrecht: D. Reidel Publishing Company, 119–121.

Dillard, Annie. February 1974. "Sight Into Insight." *Harper's Magazine* 248, 39–46.

Dobrzycki, Jerzy. 1973. "Nicolaus Copernicus—His Life and Work." *The Scientific World of Copernicus.* Bienkowska, Barbara, ed. Boston: D. Reidel Publishing Company, 13–37.

Drake, Stillman. 1957. *Discoveries and Opinions of Galileo.* Garden City, New York: Doubleday Anchor Books.

———. 1970. *Galileo Studies.* Ann Arbor: University of Michigan Press.

———. 1992. *Galileo.* New York: Oxford University Press.

Dreyer, J. L. E. 1890. *Tycho Brahe.* Edinburgh: Adam & Charles Black.

———. 1953 (reprint). *A History of Astronomy from Thales to Kepler.* New York: Dover Publications.

Dryden, John, trans. 1957. *Plutarch: The Lives of the Noble Grecians and Romans.* New York: Modern Library.

Dyson, F. W. 1915. "Measurement of the Distances of the Stars." *Observatory* 38, 249–254, 292–299.

Eichhorn, Heinrich. 1974. *Astronomy of Star Positions.* New York: Frederick Ungar Publishing Company.

Einstein, Albert, and Infeld, Leopold. 1938. *The Evolution of Physics.* New York: Simon & Schuster.

Ellis, William S. 1998. *Glass.* New York: Avon Books.

'Espinasse , Margaret. 1956. *Robert Hooke.* London: William Heinemann, Ltd.

Evans, David S. 1967. "Historical Notes on Astronomy in South Africa." *Vistas in Astronomy* 9, 265–282.

———. 1988. *Under Capricorn: A History of Southern Hemisphere Astronomy.* Philadelphia: Adam Hilger.

Evans, James. 1998. *The History and Practice of Ancient Astronomy.* New York: Oxford University Press.

Ferguson, Kitty. 1999. *Measuring the Universe: Our Historic Quest to Chart the Horizons of Space and Time.* New York: Walker and Company.

Fernie, J. D. 1975. "The Historical Search for Stellar Parallax." *Journal of the Royal Astronomical Society of Canada* 69, 153–161, 222–239.

———. 1976. *The Whisper and the Vision: Voyages of the Astronomers.* Toronto: Clarke, Irwin and Company.

Ferris, Timothy. 1988. *Coming of Age in the Milky Way.* New York: William Morrow and Company.

Fine, Gerald J. September 1991. "Glass and Glassmaking." *Journal of Chemical Education* 68, 765–768.

Forbes, Eric C., ed. 1995. *The Correspondence of John Flamsteed, the First Astronomer Royal.* Philadelphia: Institute of Physics.

Fricke, Walter. 1985. "Friedrich Wilhelm Bessel (1784–1846)." *Astrophysics and Space Science* 110, 11–19.

"The Galilean Satellites 2,000 Years Before Galileo." February 1982. *Sky & Telescope* 63, 145.

Galilei, Galileo. 1964. *Dialogue on the Great World Systems.* Salusbury translation, with introduction by Giorgio de Santillana. Chicago: University of Chicago Press.

———. 1987. *Sidereus Nuncius* (Facsimile of British Library copy). Alburgh, England: Archival Facsimiles Ltd.

———. 1989. *Sidereus Nuncius, or The Sidereal Messenger.* Van Helden, Albert, trans. Chicago: University of Chicago Press.

"Galileo Saw Neptune." November 1980. *Sky & Telescope* 60, 363.

Gamow, George. 1961. *One Two Three . . . Infinity.* New York: Viking Press.

Gause, Fritz. 1968. *Die Geschichte der Stadt Königsberg in Preussen.* Köln: Böhlau-Verlag.

Gavine, David. January 1998. "Thomas Henderson 1798–1844." *Scottish Astronomers Group Magazine.* [star-www.st-and.ac.uk/~fv/sag/dave2.htm]

Gesellschaft "Union" zu Bremen. 1890. *Bessel als Bremer Handlungslehrling.* Bremen: J. Kühtmann's Buchhandlung.

Gill, David. 1913. *A History and Description of the Royal Observatory, Cape of Good Hope.* London: His Majesty's Stationery Offices.

Gillespie, C. C. 1972. *Dictionary of Scientific Biography.* New York: Charles Scribner's Sons.

Gingerich, Owen. April 1973. "The Foundation of Modern Science." *Courier,* 10–13.

———. December 1977. "Tycho Brahe and the Great Comet of 1577." *Sky & Telescope* 54, 452–458.

———. May 1981. "Great Conjunctions, Tycho, and Shakespeare." *Sky & Telescope* 61, 394–395.

———. August 1982. "Dreyer and Tycho's World System." *Sky & Telescope* 64, 138–140.

———. October 1984(a). "Herschel's 1784 Autobiography." *Sky & Telescope* 68, 317–319.

———. December 1984(b). "Galileo and the Phases of Venus." *Sky & Telescope* 68, 520–522.

———. 1992. *The Great Copernicus Chase and Other Adventures in Astronomical History.* Cambridge, MA: Sky Publishing Corporation.

———. March 1993(a). "How Galileo Changed the Rules of Science." *Sky & Telescope* 85, 32–36.

———. 1993(b). *The Eye of Heaven: Ptolemy, Copernicus, Kepler.* New York: American Institute of Physics.

Gladwell, Malcolm. 2000. *The Tipping Point: How Little Things Can Make a Big Difference.* New York: Little, Brown.

Gould, William L. March 1989. "Small-Scale Telescope/Joseph Fraunhofer 1786–1826." *Sky & Telescope* 77, 250–251.

Grant, Robert. 1852. *History of Physical Astronomy.* London: Henry G. Bohn.

Hamel, Jürgen. 1984. *Friedrich Wilhelm Bessel.* Leipzig: BSB B.G. Teubner Verlagsgesellschaft.

Hartner, Willy. 1969. "Galileo's Contribution to Astronomy." *Vistas in Astronomy* 11, 31–43.

Harwit, Martin. 1984. *Cosmic Discovery.* Cambridge, MA: Massachusetts Institute of Technology Press.

Heath, Thomas L. 1956. *The Thirteen Books of Euclid's Elements.* New York: Dover Publications.

———. 1966. *Aristarchus of Samos: The Ancient Copernicus.* Oxford: Clarendon Press.

———. 1990. "The Sand-Reckoner," in *The Works of Archimedes Including the Method.* Chicago: Encyclopaedia Britannica, Inc.

———. 1991. *Greek Astronomy.* New York: Dover Publications.

Henderson, Thomas. 1839(a). "On the Parallax of Alpha Centauri." *Memoirs of the Royal Astronomical Society* 11, 61–68.

———. 1839(b) "On the Parallax of Sirius." *Memoirs of the Royal Astronomical Society* 11, 239–248.

Herrman, Dieter. 1984. *The History of Astronomy from Herschel to Hertzsprung.* Cambridge, England: Cambridge University Press.

Herschel, John F. W. 1829. "Address to Members (presentation of Gold Medal to Bessel)." *Monthly Notices of the Royal Astronomical Society* 1, 110–113.

———. 1842. "Address Delivered at the Annual General Meeting of the Royal Astronomical Society, February 12, 1842, on Presenting the Honorary Medal to M. Bessel." *Memoirs of the Royal Astronomical Society* 12, 442–454.

Herschel, William. 1781. "Account of a Comet." *Philosophical Transactions* 71, 492–501.

———. 1782. "On the Parallax of the Fixed Stars." *Philosophical Transactions* 72, 82–111.

———. 1803/1804. "Account of the Changes that have happened during the last Twenty-five Years, in the relative Situation of Double-stars; with an Investigation of the Cause to which they are owing." *Philosophical Transactions* 93 (1803), 339–382; 94 (1804), 353–384.

———. 1912. *The Scientific Papers of Sir William Herschel.* Dreyer, J. L. E., ed. London: Royal Society and Royal Astronomical Society.

Hevelius, Johannes. 1967. *Selenographia Sive Lunae Descripto* (Facsimile of first edition, 1647). New York: Johnson Reprint Corporation.

Hoffleit, Dorrit. 1949. "The Quest for Stellar Parallax." *Popular Astronomy* 57, 259–273.

———. 1992. *Astronomy at Yale 1701–1968.* New Haven, CT: Connecticut Academy of Arts and Sciences.

Hogben, Lancelot. 1937. *Mathematics for the Million.* New York: W. W. Norton.

Holden, Edward S. 1881. *Sir William Herschel: His Life and Works.* New York: Charles Scribner's Sons.

Hooke, Robert. 1674. *An Attempt to Prove the Motion of the Earth from Observations.* London: Royal Society.

Hoskin, Michael. 1963. *William Herschel and the Construction of the Heavens.* London: Oldbourne Book Co., Ltd.

———. 1966. "Stellar Distances: Galileo's Method and Its Subsequent History." *Indian Journal of History of Science* 1, 22–29.

———. 1982. *Stellar Astronomy.* Chalfont St. Giles, England: Science History Publications.

Hoskin, Michael, ed. 1997. *The Cambridge Illustrated History of Astronomy.* Cambridge, England: Cambridge University Press.

Howse, Derek. July 1970. "Restoration at Greenwich Observatory." *Sky & Telescope* 40, 3–9.

———. 1975. *Greenwich Observatory, Volume 3: The Buildings and Instruments*. London: Taylor and Francis Ltd.

———. 1995. *Francis Place and the Early History of the Greenwich Observatory*. New York: Science History Publications.

Hoyle, Fred. 1973. *Nicolaus Copernicus: An Essay on His Life and Work*. New York: Harper and Row.

Huffer, C. M. December 1946/ January 1947. "The Astronomy of Tycho Brahe." *Sky & Telescope* 6 (December 1946), 3–5; 6 (January 1947), 9–11.

Humberd, Charles D. 1937. "Tycho Brahe's Island." *Popular Astronomy* 45, 118–125.

Hunter, A., and Martin, E. G. 1956. "Fifty Years of Trigonometrical Parallaxes." *Vistas in Astronomy* 2, 1023–1030.

Ionides, Stephen. 1939. *Stars and Men*. New York: Bobbs-Merrill.

Jackson, J. 1922. "Early Estimations of Stellar Distances." *Observatory* 45, 341–352.

———. 1956. "The Distances of the Stars: A Historical Review." *Vistas in Astronomy* 2, 1018–1022.

James, Peter, and Thorpe, Nick. 1994. *Ancient Inventions*. New York: Ballantine Books.

Keynes, Geoffrey. 1960. *A Bibliography of Dr. Robert Hooke*. Oxford: Clarendon Press.

King, Henry C. 1955. *The History of the Telescope*. New York: Dover Publications.

Klein, Morris. 1959. *Mathematics and the Physical World*. New York: Thomas Y. Crowell Company.

Koestler, Arthur. 1989. *The Sleepwalkers: A History of Man's Changing Vision of the Universe*. New York: Penguin Books.

Kolb, Rocky. 1996. *Blind Watchers of the Sky*. Reading, MA: Addison-Wesley.

Kopal, Zdenek. 1973. "Foreword." In *The Scientific World of Copernicus*. Bienkowska, Barbara, ed. Boston: D. Reidel Publishing Company, vii–xii.

———. 1985. "Friedrich Wilhelm Bessel—an Appreciation." *Astrophysics and Space Science* 110, 3–10.

Koretz, Jane F., and Handelman, George H. July 1988. "How the Human Eye Focuses." *Scientific American* 259, 92–99.

Kovalevsky, Jean. May 1990. "Astronomy from Earth and Space." *Sky & Telescope* 79, 493–497.

Krisciunas, Kevin. 1978. "A Short History of Pulkova Observatory." *Vistas in Astronomy* 22, 26–37.

Krupp, E. C. December 1996. "Observing the Occasion [Tycho Brahe]." *Sky & Telescope* 92, 68–69.

Kuhn, Thomas S. 1979. *The Copernican Revolution: Planetary Astronomy in the Development of Western Thought*. Cambridge, MA: Harvard University Press.

Kurinsky, Samuel. 1991. *The Glassmakers.* New York: Hippocrene Books.

Labitzke, P. 1935. "Die Königsberger Sternwarte." *Die Himmelswelt* 45, 9–13.

Laurie, P. S. 1956. "Flamsteed's Well." *Observatory* 76, 24–25.

Learner, Michael. 1981. *Astronomy through the Telescope.* New York: Van Nostrand Reinhold.

Leggat, Robert. 2000. *A History of Photography.* [www.rleggat.com/photohistory/]

Leitch, Alexander. 1978. *A Princeton Companion.* Princeton, NJ: Princeton University Press.

Ley, Willy. 1963. *Watchers of the Skies.* New York: Viking.

Lodge, Oliver. 1960. *Pioneers of Science.* New York: Dover Publications.

Lovi, George. January 1985. "The Distance Dilemma." *Sky & Telescope* 69, 45–46.

———. September 1988. "An Anniversary for a Special Star [61 Cygni]." *Sky & Telescope* 76, 275–276.

Lubbock, Constance A. 1933. *The Herschel Chronicle: The Life Story of William Herschel and His Sister Caroline Herschel.* Cambridge, England: Cambridge University Press.

Main, R. 1842. "On the Present State of Our Knowledge of the Parallax of the Fixed Stars." *Memoirs of the Royal Astronomical Society* 12, 1–60.

Marly, Pierre. 1988. *Spectacles and Spyglasses.* Paris: Editions Hoebeke.

McCrea, W. H. 1963(a). "James Bradley, 1693–1762." *Quarterly Journal of the Royal Astronomical Society* 4, 38–40.

———. 1963(b) "The Significance of the Discovery of Aberration." *Quarterly Journal of the Royal Astronomical Society* 4, 41–43.

McCutcheon, Robert A. December 1994/January 1995. "Sunset on Pulkova." *Air & Space Magazine* 9, 38–45.

Miller, Donald T. January 2000. "Retinal Imaging and Vision at the Frontiers of Adaptive Optics." *Physics Today* 53, 31–36.

Moore, Patrick. 1989. *Astronomers' Stars.* New York: W. W. Norton.

———. 1994. *The Great Astronomical Revolution: 1534–1687 and the Space Age Epilogue.* Chichester, England: Albion Publishing.

Moore, Patrick, and Collins, Peter. 1977. *The Astronomy of Southern Africa.* London: Robert Hale and Company.

Morrison, Philip. February 1998. "The Star Mapper." *Scientific American* 278, 100–102.

Morrison, Philip, and Morrison, Phylis. May 2000. "Netting the Deep Sky." *Scientific American* 282, 116–118.

Motz, Lloyd, and Duveen, Anneta. 1977. *Essentials of Astronomy,* 2nd ed. New York: Columbia University Press.

Motz, Lloyd, and Weaver, Jefferson H. 1995. *The Story of Astronomy.* New York: Plenum Press.

Murray, C. A. 1988. "The Distances to the Stars." *Observatory* 108, 199–217.

Neugebauer, O. 1975. *A History of Ancient Mathematical Astronomy*. New York: Springer Verlag.

Nichols, Richard. 1999. *Robert Hooke and the Royal Society*. Sussex, England: The Book Guild, Ltd.

Nielsen, Axel V. 1968. "Ole Roemer and His Meridian Circle." *Vistas in Astronomy* 10, 105–112.

North, John. 1995. *Norton History of Astronomy and Cosmology*. New York: W. W. Norton.

Olbers, Wilhelm. 1852. *Briefwechsel Zwischen W. Olbers und F. W. Bessel*. Leipzig: Avenarius and Mendelssohn.

Oriti, Ronald A., and Starbird, William B. 1977. *Introduction to Astronomy*. Encino, CA: Glencoe Press.

Panek, Richard. 1998. *Seeing and Believing: How the Telescope Opened Our Eyes and Minds to the Heavens*. New York: Viking.

Pannekoek, A. 1969. *A History of Astronomy*. New York: Barnes and Noble.

Park, David. 1997. *The Fire within the Eye: A Historical Essay on the Nature and Meaning of Light*. Princeton, NJ: Princeton University Press.

Paterson, E. Russell. February 1957. "Robert Hooke." *Sky & Telescope* 16, 179–180.

Payn, Howard. 1914. "The Well of Eratosthenes." *Observatory* 37, 286–288.

Pearson, William. 1824–1829. *An Introduction to Practical Astronomy*. London: Longman, Hurst.

Pederson, Olaf. April 1973. "The Making of a New Universe." *Courier*, 14–18.

Pedoe, Daniel. 1976. *Geometry and the Liberal Arts*. New York: St. Martin's Press.

Perryman, Michael. June 1998. "The Hipparcos Astrometry Mission." *Physics Today* 51, 35–43.

———. June 1999. "Hipparcos: The Stars in Three Dimensions." *Sky & Telescope* 97, 40–50.

Peterson, Ivars. December 18/25, 1999. "Gravity Tugs at the Center of a Priority Battle." *Science News* 156, v.

The Principles of Rangefinding. 1911. Glasgow: Barr and Stroud.

Range Finder M7 (War Department Technical Bulletin TB 9-585-1). 1944. Washington, DC: War Department.

Range Finder M9 (War Department Technical Manual TM 9-585). 1943. Washington, DC: War Department.

Repsold, Johann A. 1908. *Zur Geschichte der Astronomischen Messwerkzeuge von Purbach bis Reichenbach, 1450 bis 1830*. Leipzig: Verlag von Wilhelm Engelmann.

———. 1919. "Friedrich Wilhelm Bessel." *Astronomische Nachrichten* 210, 161–214.

Rigaud, S. P. 1832. *Miscellaneous Works and Correspondence of the Rev. James Bradley.* Oxford: Oxford University Press.

Roberts, Royston M. 1989. *Serendipity: Accidental Discoveries in Science.* New York: John Wiley and Sons.

Rogers, Eric M. 1982. *Astronomy for the Inquiring Mind.* Princeton, NJ: Princeton University Press.

Rogers, Frances, and Beard, Alice. 1948. *5000 Years of Glass.* New York: J. B. Lippincott Company.

Ronan, Colin. February 1964. "Galileo Galilei—1564–1642." *Sky & Telescope* 27, 72–78.

———. March 1981. "William Herschel and his Music." *Sky & Telescope* 61, 195–204.

Ronchi, Vasco. 1967. "The General Influence of the Development of Optics in the Seventeenth Century on Science and Technology." *Vistas in Astronomy* 9, 123–133.

Rosen, Edward. 1956. "The Invention of Eyeglasses." *Journal for the History of Medicine and Allied Sciences* 11, 13–46, 183–218.

———. 1971. *Three Copernican Treatises.* New York: Octagon Books.

———. trans. Dobrzycki, Jerzy, ed. 1978. *Nicolaus Copernicus: On the Revolutions.* Baltimore: Johns Hopkins University Press.

———. June 1981. "Render Not Unto Tycho That Which Is Not Brahe's." *Sky & Telescope* 61, 476–477.

Rosenthal, J. William. 1996. *Spectacles and Other Vision Aids: A History and Guide to Collecting.* San Francisco: Norman Publishing.

Roth, Gunter D. 1976. *Joseph von Fraunhofer.* Stuttgart: Wissenschaftliche Verlagsgesellschaft mbH.

Roy, A. E., and Clark, D. 1977. *Astronomy: Principles and Practice.* New York: Crane Russell.

Sacks, Oliver. 1995. *An Anthropologist on Mars.* New York: Alfred A. Knopf.

Sagan, Carl. 1980. *Cosmos.* New York: Random House.

de Santillana, Giorgio. 1955. *The Crime of Galileo.* Chicago: University of Chicago Press.

Sarton, George. 1931. *Introduction to the History of Science.* Baltimore: Williams and Wilkins.

———. 1957. *Six Wings: Men of Science in the Renaissance.* Bloomington: Indiana University Press.

———. 1959. *A History of Science: Hellenistic Science and Culture in the Last Three Centuries B.C.* Cambridge, MA: Harvard University Press.

Schlesinger, Frank. 1927. "Some Aspects of Astronomical Photography of Precision." *Monthly Notices of the Royal Astronomical Society* 87, 506–523.

Schmeidler, F. July 1984. "Der Astronom Friedrich Wilhelm Bessel." *Nachrichten der Olbers-Gesellschaft,* no. 130. [www.rz.hs-bremen.de/planetarium/plabesse.htm]

Schweiger-Lerchenfeld, Armand von. 1898. *Atlas der Himmelskunde.* Vienna: A. Hartleben's Verlag.

Shapley, Harlow. 1960. *Source Book in Astronomy 1900–1950*. Cambridge, MA: Harvard University Press.

Shapley, Harlow, and Howarth, H. E. 1929. *A Source Book in Astronomy*. New York: McGraw-Hill.

Sidgwick, J. B. 1954. *William Herschel: Explorer of the Heavens*. London: Faber and Faber, Ltd.

Sime, J. 1900. *William Herschel and His Work*. Edinburgh: T. & T. Clark.

Simonsen, Erik. February 1974. "A Visit to Tycho Brahe's Observatory." *Sky & Telescope* 47, 86–88.

Simplex. 1916. *One-Man Range Finders and How to Use Them*. London: Forster Groom and Company.

Slocum, Frederick. 1967. "Stellar Parallax." In *Starlight: What It Tells About the Stars*. Page, Thornton, and Page, Lou Williams, eds. New York: Macmillan.

Smart, W. M. 1950. *Some Famous Stars*. London: Longman Green.

Sobel, Dava. 1999. *Galileo's Daughter*. New York: Walker & Company.

Spencer-Jones, Harold. 1941. "The Solar Parallax: A Coordinated International Measure of a Fundamental Constant." *Observatory* 64, 99–104.

Stewart, Albert B. March 1964. "The Discovery of Stellar Aberration." *Scientific American* 210, 100–108.

Stille, Alexander. May 8, 2000. "Resurrecting Alexandria." *The New Yorker*, 90–99.

Strand, K. A. January 1942. "The Double Star System 61 Cygni." *Sky & Telescope* 1, 6–8.

Struve, Otto. November/December 1956. "The First Determination of Stellar Parallax." *Sky & Telescope* 16 (November 1956), 9–12; 16 (December 1956), 69–72.

———. 1959. "The First Stellar Parallax Determination." In *Men and Moments in the History of Science*. Evans, H. M., ed. Seattle: University of Washington Press.

Suter, Rufus. November 1951. "Galileo in Padua." *Sky & Telescope* 11, 3–4.

Thiel, Rudolph. 1957. *And There Was Light: The Discovery of the Universe*. New York: Alfred A. Knopf.

Thoren, Victor. 1990. *The Lord of Uraniborg: A Biography of Tycho Brahe*. New York: Cambridge University Press.

Toomer, G. J., trans. 1984. *Ptolemy's Almagest*. New York: Springer Verlag.

Turner, Herbert Hall. 1963. *Astronomical Discovery*. Berkeley: University of California Press.

Turon, Catherine. July 1997. "From Hipparchus to Hipparcos." *Sky & Telescope* 94, 28–34.

Unsöld, Albrecht. 1969. *The New Cosmos*. New York: Springer Verlag.

van de Kamp, Peter. 1985. "Friedrich Wilhelm Bessel: 1784, July 22–1846, March 17." *Astrophysics and Space Science* 110, 103–104.

van Helden, Albert. 1977. *The Invention of the Telescope*. Philadelphia: American Philosophical Society.

———. 1985. *Measuring the Universe: Cosmic Dimensions from Aristarchus to Halley.* Chicago: University of Chicago Press.

Warner, Brian. 1995. *Royal Observatory, Cape of Good Hope, 1820–1831: The Founding of a Colonial Observatory.* Boston: Kluwer Academic Publishers.

Warner, Brian, and Warner, Nancy. 1984. *Maclear and Herschel: Letters and Diaries at the Cape of Good Hope, 1834–1838.* Cape Town: A. A. Balkema.

Wattenberg, D. 1929(a). "Die Alte Sternwarte in Lilienthal." *Das Weltall* 28, 123–125.

———. 1929(b). "Bessel als Bremer Kaufmannslehrling." *Das Weltall* 29, 8–11.

———. 1933. "Wilhelm Olbers." *Die Himmelswelt* 44, 176–184.

———. 1934. "Friedrich Wilhelm Bessel." *Die Himmelswelt* 44, 124–135.

Whitney, Charles. 1971. *The Discovery of the Galaxy.* New York: Alfred A. Knopf.

Williams, Henry Smith. 1915. *Modern Warfare.* New York: Hearst's International Library.

Williams, M. E. W. 1979. "Flamsteed's Alleged Measurement of Annual Parallax for the Pole Star." *Journal for the History of Astronomy* 10, 102–116.

———. 1981. *Attempts to Measure Annual Stellar Parallax: Hooke to Bessel.* Ph.D. thesis, Imperial College, University of London.

Wilmoth, Frances, ed. 1997. *Flamsteed's Stars.* Rochester, NY: Boydell Press.

Woolley, Richard. 1963. "James Bradley, Third Astronomer Royal." *Quarterly Journal of the Royal Astronomical Society* 4, 46–52.

Zeilik, Michael, Gregory, Stephen, and Smith, Elske V. P. 1992. *Introductory Astronomy and Astrophysics,* 3rd ed. New York: Saunders College Publishing.

Index ★★★

aberration, 160–161, 167
  chromatic, 237, 238
Accademia dei Lincei (Academy
  of the Lynxes), 125
accommodation (vision), 53
achromatic lens, 237–239, 241
acoustic ranging, 55
air compressor, 140
Airy transit circle, 221
Albireo, 221
Alcor, 132
Aldebaran, 37, 204
Alexander I (czar of Russia),
  242, 253–255
Alexandria, Egypt, 2
Alexandria Library, 2, 29
Alexandrian Museum, 2
Algol, 138
Allegheny Observatory, 274, 275
Almagest (Ptolemy), 23–24, 27,
  28, 30, 39
Alpha Centauri, 194, 277
  parallax of, 200, 204–205,
    263, 264
  position of, 203, 204
  proper motion of, 204
Alpha Herculis, 219
Altair, 254
altazimuth mounts, 241–242
Anaximander, 4
aperture, 172
Apollonius of Perga, 25
apparent double stars, 260
Aquinas, Thomas, 30
Arago, François, 218, 272
Archer, Frederick Scott, 273
Archimedean Engine, 146–148
Archimedean screw, 10
Archimedes, 10–15
arcminute, 68

arcsecond, 68
Arcturus, 204
Arecibo Radio Observatory,
  193–194
Aristarchus of Samos, 1, 6–10,
  12–17, 20, 45, 47, 186, 279
Aristotle, 4, 5, 14, 20, 24, 25, 28,
  30, 89
Ashbrook, Joseph, 88
asteroids, 65
astrometry, 214
astronomical unit, 60, 61, 65–66
astronomy
  and photography, 272–274
  positional, 226
  precision, 161–165, 226
  southern hemisphere, 195–205
atmosphere
  and parallax measurement,
    275–276
atmospheric refraction, 144–145
Aubrey, John, 139, 140, 142, 143,
  195
Averroes, 41

Bach, Johann Sebastian, 151
Bacon, Roger, 95, 109
Badovere, Jacques, 119
Banks, Joseph, 185
Barisino, Tommaso, of Modena,
  94, 100
Baronius, Cardinal, 113
Bass, George, 238
Bayer, Johannes, 136
Bessel, Friedrich Wilhelm, 165,
  206–221, 225, 236, 242,
  243, 246, 248, 250–251,
  254, 255, 258–267, 269,
  271, 272, 277, 280, 284
Betelgeuse, 270

Biela's comet, 203, 211
binary stars, see double stars
binary-star systems, masses of,
    278
binocular range finder, 55, 56
Bird, John, 164, 226
black holes, 187–188
Blake, William, 95
Bodin, Jean, 43
Bond, William Cranch, 273
Boyle, Charles, Earl of Orrery,
    33
Boyle, Robert, 138
Bradley, James, 148–165, 168,
    186–187, 217, 226, 270,
    280, 284
Brahe, Tycho, 47, 60–61, 74, 75,
    77–93, 118, 129–130, 155,
    279, 284
Brinkley, John, 219
Browne, Sir Thomas, 49
Bunsen, Robert, 240
Burney, Fanny, 184
Busch, Georg, 84

Calandrelli, Giuseppe, 218
calculator, mechanical, 141
Canopus, 270
Cape of Good Hope, Royal
    Observatory, see Royal
    Cape Observatory
Cassini, Giovanni Domenico, 61
Cassini, J. D., 226–227
celestial navigation, 209
Cesi, Prince Federico, 125
charged-coupled devices
    (CCDs), 275
Charles II (king of England),
    195–196
Chésaux, Philippe Loys de, 167
Chesterton, G. K., 19
Christian (king of Denmark), 92
chromatic aberration, 237, 238
Churchill, Winston, 223
circle-style telescope, 220, 221
Clark, Alvan Graham, 267
Cleanthes of Assus, 16–17

Clerke, Agnes, 4–5, 152, 195, 269,
    270, 271
comet panic of 1832, 211
comets, 89, 180, 246
    see also individual comets,
    e.g., Biela's comet
Commentariolus (Copernicus),
    39–40
compound lens, 238
Comte, Auguste, 241
concave lenses, 100, 101,
    106–109
convex lenses, 99, 101, 106–109
Cook, Captain James, 64
Copernican system, 45–46, 60,
    91, 125–129, 144–148, 161
Copernicus, Nicolaus, 35–47,
    279
Cosimo II de' Medici, 120, 124
Cremonini, Cesare, 118
Cristofori, Bartolomeo, 151
crown glass, see soda-lime glass
cullet, 105
Cutler, Sir John, 141

Daguerre, Louis Jacques, 272
daguerreotypes, 272, 273
Davis, Herman, 274
De la Rue, Warren, 273, 274
De Revolutionibus (Copernicus),
    42–46, 126
deferent (of an orbit), 5, 25–29
Demisiani, Giovanni, 125
Democritus, 96
Deneb, 221, 254, 270
depth perception, 50, 51
Dialogo . . . (Galileo), 127–128,
    130
Dillard, Annie, 51
distance, measuring with
    parallax, 51–53
distance perception, 50–52, 54
Dixon, Jeremiah, 64
Dollond, John, 238–239, 258
Dollond telescopes, 199,
    249–250
Dorpat Observatory, 249

Dorpat refractor, see Great Refractor

Double Interferometer for Visual Astronomy (DIVA), 277–278, 284

double stars, 132, 168, 179, 185–186, 219–220, 253–254

apparent double, 260

and parallax, 187–189

double-star method of stellar parallax, 131–132, 173, 187

Drake, Stillman, 115

Dreyer, J. L. E., 135

Dryden, John, 143

Earth, spinning of, 5–6

Earth-centered model of the universe, see geocentric model of the universe

Earth-Sun distance (astronomical unit), 60, 61, 65–66

eclipses

in 189 B.C., 59

in 585 B.C., 4

in 1560, 78

of Aldebaran, 204

Einstein, Albert, 3

Elements (Euclid), 24

Elkin, W. Lewis, 270

Eltamin, see Gamma Draconis

Enceladus, 185

Encke's comet, 203

England, glass manufacture in, 232

epicycle (of an orbit), 5, 25–29

equant, 27–28

equatorial mounts, 241–242

Eros (asteroid), 65–66

Euclid, 24

Eudoxus of Cnidus, 5

Euripides, 113

European Space Agency (ESA), 276–278

Evans, David S., 62

Everest, Sir George, 226

eye lens, 97–100

eyepiece lens, 106–107

eyepiece micrometer, 187

eye-to-eye baseline, 54–56

Fallows, Fearon, 194, 196–201

Ferguson, James, 176

Fitz, Henry, 273

Fizeau, Armand Hippolyte, 273

Flamsteed John, 135, 139–140, 147–148, 162, 284

Flat-Earth Society, 4

flint glass (lead crystal), 231–232

Flying Star, see 61 Cygni

flying stars, 270

Foscarini, Paolo Antonio, 43

Foucault, Jean, 273

Fraunhofer, Joseph, 222–225, 227–233, 236–237, 239–243, 246, 251–253, 257, 258, 280

Fraunhofer lines, 240–241

Frederick II (king of Denmark), 85

Friedrich Wilhelm III (king of Prussia), 216, 227

frit, 105

Full-sky Astrometric Mapping Explorer (FAME), 278, 284

Galilei, Galileo, 38, 43, 44, 47, 93, 112–133, 142, 173, 188, 279, 280

Galilei, Vincenzio, 114

Galle, Johann, 123, 266

Gamma Draconis (Eltamin), 136–138, 145–148, 154–155, 156–157, 158, 161, 165, 168

Gamow, George, 14, 53

Gascoigne, William, 155

Gauss, Carl Friedrich, 211

Gelon II (king of Syracuse), 12, 15

geocentric model of the universe, 3–9, 16–18, 21, 23

Ptolemaic version of, 23–31

George III (king of England), 181, 182, 184
Germany, optical workshops in, 227
giant stars, 278
Gibbon, Edward, 207
Giese, Bishop, 44
Gill, David, 196, 269–270
Gingerich, Owen, 24, 165
Gladwell, Malcolm, 106
glass, 101–102
    optical, 232–233, 241
glassblowing, 102
glassmaking, 102–105
    in Venice, 102–103
Global Astrometric Interferometer for Astrophysics (GAIA), 278, 284
Graham, George, 154, 157, 163, 226
gravity, 143, 171
    laws of, 189
Great Comet of 1807, 246
Great Refractor (Dorpat), 234, 242, 243, 251–253, 255, 259, 263, 264, 277
Green, Charles, 64
Greenwich mural circle, 192, 200, 221
Greenwich Observatory, 274
Gregory, David, 238
Gregory, James, 62
Gresham, Thomas, 141
Guinand, Pierre Louis, 232–233
guncotton, 273
Gutenberg, Johann, 35

Hall, Chester Moor, 238, 239
Hall, R. H., 102
Halley, Edmond, 22, 61–65, 210, 270
    and James Bradley, 153, 161
    and Hooke-Hevelius dispute, 146
    and Royal Observatory, 162
    and southern hemisphere astronomy, 195–196
    and stellar motions, 204
Halley's comet, 64, 210
Handel, Georg Friedrich, 151
Harkness, William, 65
Harriot, Thomas, 210
Harvard Observatory, 272
Helden, Albert van, 61, 101
heliocentric model of the universe, xvi, 9, 20–21
    Aristarchus model of, 1–2, 6–8, 10, 12–17
    Copernican system, 39–47
heliometer, 256–260, 269–270, 284
Helmholtz, Hermann von, 51
Henderson, Thomas, 190–191, 193, 194, 201–205, 263, 265–266, 277, 280
Henry the Navigator, 30
Heraclides of Pontus, 7
Herodotus, 3
Herschel, Alexander, 178
Herschel, Caroline, 174–176, 178–179, 181–185, 189–190
Herschel, John, xv, 185, 190, 196, 197, 258, 262, 265, 266
Herschel, William, 170–190, 195, 214, 219, 237, 270, 280
Hevelius, Johannes, 139, 145–146, 237
Hieron (king of Syracuse), 11, 12
Hipparchus, 21–23, 25, 59, 82–83, 204, 276, 279, 284
Hipparcos (High Precision Parallax Collecting Satellite), 276–277, 284
Holmes, Oliver Wendell, 184
Hooke, Robert, 134, 138–149, 151–152, 165, 280
Hooke's law (of spring action), 141
Hoskin, Michael, 30
Hubble Space Telescope, 276
Humboldt, Alexander von, 216, 267
Huth, Johann, 248, 249

Huygens, Christiaan, 166–167
Hven, Denmark, 85–88
Hyades, 277

Ibn al-Haitham (Alhazen), 95
Ibn as-Shatir, 45
Ikeya-Seki comet, 245
inertia, 6–7, 143
Infeld, Leopold, 3
infrared light, 186
International Astronomical
       Union, 276
inverse square law, 143
iris diaphragm, 141
Islamic science, 29–30

Jagiellonian University, 36
Janssen, Sacharias, 110
Johnson, Samuel, 95
Jupiter, 46
    moons of, 122–124
    orbit of, 60

Kepler, Johannes, 28, 47, 83, 89,
       92–93, 117–118, 124, 142
Kirchhoff, Gustav, 240
Kitchiner, William, 223
Koestler, Arthur, 19, 92
Königsberg heliometer, 256,
       257–260
Königsberg Observatory,
       216–217
Kuffner Observatory, 270

Lacaille, Nicolas Louis de, 60,
       196
Lambert, J. H., 167
Le Gentil, Guillaume, 64
Leacock, Stephen, 171
lead crystal, see flint glass
Leander McCormick
       Observatory, 274
lens design, 239
lens making, 231–233, 241
lens pairs, 106–109
lenses, 96–97, 106–109
    achromatic, 237–239, 241

compound, 238
concave, 100, 101, 106–109
convex, 99, 101, 106–109
of the eye, 97–100
eyepiece, 106–107
objective, 106–107, 237
telescope, 231–232
Ley, Willy, 90
Liebherr, Joseph, 229–230, 233
Lieven, Prince, 251
Lindenau, Baron von, 218
Lipperhey, Hans, 109–110
Luther, Martin, 43

Macclesfield, Earl of, 152
Maclear, Thomas, 193, 202
Maddox, Richard, 274
Mahler (telescope maker), 254,
       269
Mann, Sir Horace, 158
Marcellus (Roman general), 10,
       15
Mars, 46, 93
    moons of, 142
    orbit of, 60
    parallax of, 60–61, 203
Maskelyne, Neville, 64, 179, 180,
       187
Mason, Charles, 64
Mathematical-Mechanical
       Institute of Munich,
       229–233
Mathieu, Claude-Louis, 218
Maximilian Joseph, Prince, 224,
       225
Meadows, William, 203, 205, 263
Medicean Stars, 124
Medici family, 120, 124
Mercury, 46
    motion of, 20
    parallax of, 62
    transit of, 203
mercury amalgam, 141
meridian, 162–163
Merz, Georg, 233, 254, 269
Metius, Jacob, 110
Michell, John, 167, 187–188

Micrographia (Hooke), 141
Milky Way, 122, 186
Milton, John, 33
Mimas, 185
Mizar-Alcor pair, 132
Molyneux, Samuel, 154–158
Monod, Jacques, 151
Moon
    distance from Earth, 59
    mountains on, 185
    parallax of, 59–60, 203
    surface of, 122, 273
motion
    laws of, 189, 204
    retrograde, 25–27, 46
Mount Wilson Observatory, 274
mural circle telescopes, 204
    at Greenwich, 192, 200, 221
mural quadrant, 163, 164, 200
myopia, 100–101

Napoleon, 216
NASA, 277–278
Near Earth Asteroid Rendezvous
    spacecraft, 66
nearsightedness, *see* myopia
nebulae, 186
Neptune, 123, 266
Newton, Isaac, 23, 47, 92, 127
    on James Bradley, 152, 153
    on gravity, 171
    and Robert Hooke, 138, 139,
        141, 142–143
    laws of gravity and motion
        by, 189
    on nutation, 161
    photometric method of stel-
        lar distances, 165–166
    and solar spectrum, 239–240
    and star gauging, 186
    and stellar brightness, 167
    telescopes of, 177, 238
Niggl, Joseph, 228, 230, 231
Novara, Domenico Maria da, 37
nutation, 161

Oberon, 185

objective lens, 106–107, 237
obsidian, 102
Olbers, Wilhelm, 210–212, 219,
    261
opposition (of planet), 60
optical glass, 232–233, 241
optics, 95–96
orbits, geostationary, 276
orrery, 32–35
Osiander, Andreas, 42–43
Ovid, 171

Palermo circle, 220, 221
Pallas, 211
Palmerston, Lord, 196
parallax, xii, 8–9, 49–50
    and measuring distance,
        51–53
    and solar system, 58–66
    and surveying, 56–58, 72
    see also stellar parallax
parallax sensing, 53, 54
parallax shift, 52, 156–157
Parsons, William, Third Earl of
    Rosse, 235
Paul V, Pope, 124, 126
Peckham, John, 95
pendulum clock, 116
Pepys, Samuel, 138, 141
Perryman, Michael, 279
Petrarch, 98
Peurbach, Georg, 37
photography
    and astronomy, 272–274
    and stellar parallax, 274
photometric method of stellar
    distances, 165–166
photons, xi, 275–276
Piazzi, Giuseppe, 182, 218, 220,
    221
Pitt, Mary, 184
Place, Francis, 147
planetary motion, 4–5, 19, 20,
    24–25, 27–29, 41, 45–46
planets
    brightness of, 8
    elliptical orbits of, 47, 93

planets (*continued*)
  orbits of, 23–24, 27–29, 46, 60
  parallaxes of, 60
  positions of, 90
  retrograde motion of, 46
Plato, 3
Pleiades star cluster ("Seven
  Sisters"), 122, 277
Pliny the Elder, 22, 102
Plutarch, 10, 11, 15
Pluto, 70–71
polar axis (of telescope mount),
  241
Pond, John, 219
positional astronomy, 226
Pound, Reverend James, 153, 154
Pratensis, Johannes, 84
precession (of the Earth), 23
precision astronomy, 226
presbyopia, 97–98
Pritchard, Charles, 274
Procyon, 270
proper motion, 204, 270
Ptolemaic geocentric model of
  the universe, 23–31, 37,
  38–39, 41, 44–46, 91, 125,
  127, 145
Ptolemy, Claudius, 23–25, 27–31,
  279
Ptolemy I Soter, 2
Ptolemy III, 2
Pulkova Observatory, 248
Pythagoras of Samos, 4

quadrant, 63, 81, 89
  mural, 163, 164, 200
quartz, 101, 102

rainbow, 240
Ramsden, Jesse, 48, 220–221,
  226–227, 230
range finders, 54–56, 72
Ravenscroft, George, 231
reflector telescopes, 172,
  177–179, 180, 237, 238,
  275
  Herschel's, 183–185

Newton's, 143
Yerkes Observatory, 268
refraction, atmospheric, 144–145
refractor telescopes, 177, 237,
  239
  Great Refractor (Dorpat),
    234, 242, 243, 251–253,
    255, 259, 263, 264, 277
Reichenbach, Georg Friedrich
  von, 229–230, 233, 243
Reichenbach circle, 218, 221
Reinhold, Erasmus, 40
Repsold heliometers, 269
retrograde motion, 25–27
  of planets, 46
Rheticus, Georg Joachim, 40–43
Ricci, Ostilio, 115
Richer, Jean, 61
Rigel, 270
Rittenhouse, David, 33–35
Rittenhouse orrery, 32–35
Rivalto, Giordano da, 98
Robinson, Edwin Arlington, 269
Römer, Ole, 163
Ronchi, Vasco, 98
Rosen, Edward, 123
Royal Cape Observatory,
  196–205, 274
Rudolph II (German emperor),
  92
Rutherfurd, Lewis Morris,
  273–274

Sacks, Oliver, 50–51
Sagredo, Giovanfrancesco, 114
Saint-Exupéry, Antoine de, 33
The Sand-Reckoner
  (Archimedes), 13, 14
Santillana, Giorgio de, 127, 128
Sarpi, Paolo, 118–119
satellites (stars), 124
Saturn, 46
  brightness of, 167
  rings of, 125, 142, 185
  satellites of, 185
Schickel, Richard, 245
Schlesinger, Frank, 274, 275

Schröter, Johann Hieronymous,
212–213, 216, 218–219
Seleucus of Seliucia, 17
serendipity, 158
Shakespeare, William, xv, 86, 193
Shelley, Percy Bysshe, 75
Sidereus Nuncius (Galileo),
123–124
silica
forms of, 101
and glass, 101–102, 104
Sirius, 8, 165–167, 204, 240, 270
parallax of, 265
proper motion of, 267
61 Cygni, 221, 255, 258–260, 277
parallax of, 251, 260–264,
274
sky-coordinate system, 22–23
Smith, Robert, 177
soda (sodium carbonate or
sodium oxide), and glass-
making, 104
soda-lime glass, 104, 105, 231
solar-stellar equivalence, 166
solar system
movement of, 186
parallaxes in, 58–66
southern hemisphere astronomy,
195–205, 269–270
Space Interferometry Mission
(SIM), 278, 284
space satellites, and stellar parallax
determination, 276–277
spectacles, 94, 96, 97–101
spectral lines, 240–241
spectroscope, 240
Spina, Alessandro della, 98
spiral balance spring, 140
spring action (Hooke's law), 141
spring scale, 141
spyglass, 106–111
by Galileo, 119–122, 124, 125
stadia (measure), 15
star catalogues
by Tycho Brahe, 90, 277
by Thomas Henderson, 203
by William Herschel, 187

by Hipparchus, 21
by Giuseppe Piazzi, 221
by Wilhelm Struve, 253–254,
255, 261
star clusters, 186, 279
star formation, 279
star gauging, 186, 189
star positions, 214
Bradley's measurements of,
164–165
determination of, 226
starlight, 279
stars
apparent double, 260
double, 132, 168, 179, 185,
219–220, 253–254
giant, 278
light output of, 278, 279
movement of, 279
positions of, 22–23
white dwarf, 267
stellar brightness, and parallax,
270
stellar distances, 277, 278
Newton's photometric
method for, 165–166
stellar mass, 189, 253, 278
stellar motions, 204
stellar parallax, xii–xiii, 9–10,
16–17, 45, 47, 66–73, 91,
126, 129–133, 165, 204, 251
Bessel on, 218–220
Bradley on, 151
Brinkley on, 219
catalogues, 276–277
and charged-coupled devices
(CCDs), 275
detecting, 77, 138
determined by photography,
274–275
and double stars, 187–189
and double-star method, 173
and Earth's atmosphere,
275–276
of faint stars, 275
of Gamma Draconis,
155–157

stellar parallax (*continued*)
Henderson on, 191, 205
Herschel on, 173, 186–189
Hooke on, 143, 147, 148
measurement errors, 271
measuring, 214, 269–272,
276–278
search for, 168–169, 259, 260,
264, 279–281, 284
of 61 Cygni, 260–263
use of data from, 278–279
stellar positions, determination
of, 161
Stjerneborg observatory, 88
Struve, Jacob, 245, 246, 247
Struve, Otto, 257, 264
Struve, Wilhelm, 242–244,
246–255, 258, 260–261,
263–264, 265–266, 269,
277, 280, 284
Summer Triangle, 254
Sun
movement of, 19–20, 24–25
photographs of, 273
spectrum of, 239–241
and stellar brightness, distance, 166–167
surface of, 185
Sun-centered model of the universe, *see* heliocentric
model of the universe
sunspots, 125, 273
supernovae, 82–85
surveying, and parallax, 56–58,
72–73
Swift, Jonathan, 151

targeting, parallax-based, 58
telescope making, 236–243
telescope mirrors, 177–178
telescopes, 96, 101, 109, 125,
214–215
aperture, 172
Archimedean Engine,
146–148
Bradley's Wanstead,
157–158, 161

circle-style, 220, 221
clock-driven, 141
Dollond, 199, 249–250
Dorpat transit telescope,
249–250
Great Refractor (Dorpat),
234, 242, 243, 251–253,
255, 259, 263, 264, 277
Herschel's, 177–186
Königsberg heliometer, 256,
257–260
lenses, 231–232, 239
mounts for, 241–242
mural circle, 192, 200, 204, 221
Newtonian design, 177
reflector, 143, 172, 177–180,
183–185, 237, 238, 275
refractors, 177, 234, 237, 239,
242, 243, 277
Schröter's reflector, 212–213
and stellar parallax, 130
transit, 163–164, 199,
249–250
Yerkes Observatory, 268
zenith, 134, 154–157, 164
see also spyglass
Thales of Miletus, 3–4
theodolite, 48, 56–57
Thiel, Rudolph, 36, 235
Thomson, Thomas, 158
Titania, 185
transit circles, 218, 221
transit telescopes, 163–164, 199,
249–250
triangulation, 56–57
Troughton, Edward, 192, 200,
226
Turner, Herbert Hall, 152
Twain, Mark, 1
Tychonic system, 91–92, 125, 145

Ugone, Cardinal, Hugh of
St. Cher, 100
Ulugh Beg, 136
United States Naval Observatory,
275
universal joint, 141

Updike, John, 68
Uraniborg, 86–90, 92
Uranus, 172
  discovery of, 180–181, 185
  satellites of, 185
Urban VIII, Pope, 127, 128
Utzschneider, Joseph von, 224,
  225, 229–233

vacuum pump, 140
van Gogh, Vincent, 1
Vedel, Anders Sörensen, 78–79,
  87
Vega, 254–255, 273, 277
  parallax of, 255, 260–261,
    263, 264
Venus, 125
  motion of, 20
  orbit of, 60
  parallax of, 61–64
  transits of, 64–65
vision, 50–51
  and parallax sensing, 50, 53,
    72
Vitellius (Witelo), 95
Vivaldi, Antonio, 151

Waller, Richard, 139
Wallis, John, 142

Walpole, Horace, 158
Washington, George, 56
water glass, 104
Watson, William, 181
Watzenrode, Lucas, 36–38
Weichselberger, Philipp, 223,
  224, 228–229
Westfall, Richard S., 139
wet-collodion photography
  method, 273
wheel barometer, 140
Whipple, J. A., 273
white dwarf star, 267
white light, 240
Whitman, Walt, 207
Wilhelm IV of Hesse, 85
wind gauge, 141
Wittgenstein, Ludwig, 49
Wollaston, William Hyde, 240
Woodruff, Julia Louise, 257
Wren, Christopher, 138, 148

Yale University Observatory, 275
Yerkes Observatory, 274
  refractor, 268

zenith distance, 163
zenith telescope, 134, 154–157,
  164